9TH ANNUAL
WORLD

Maker Faire NEW YORK

M000086212

A CELEBRATION OF

INVENTION
Creativity and
CURIOSITY

INSPIRE THE FUTURE

Immerse yourself in a bounty of exhibits, stages,
interactive art, hands-on making and learning —
all showcasing the creative and experimental
minds who make, play, tinker, and hack.

SEPT 22+23

NEW YORK HALL OF SCIENCE | SAT & SUN 10AM-6PM

M

Get Tickets Today!

makerfaire.com

  @makerfaire
#wmfny18

CONTENTS

Make: **Volume 65** Oct/Nov 2018

ON THE COVER:
littleBits founder and CEO Ayah Bdeir jams on a customized Synth Kit, one of many different snap-together electronics kits offered by the company.
Photo: Mark Madeo

12

8

40

46

36

60

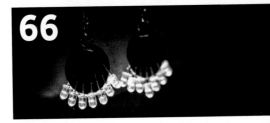

66

22

Mark Madeo, Scott Slagerman, Larry Cotton, Erin St. Blaine, Hep Svadja, Tucker Shannon, William Lambrecht

Make:

"Sharing defies the laws of physics — the more you give away the more you have."
—Adam Savage

EXECUTIVE CHAIRMAN & CEO
Dale Dougherty
dale@makermedia.com

CFO & COO
Todd Sotkiewicz
todd@makermedia.com

EDITORIAL

EDITORIAL DIRECTOR
Roger Stewart
roger@makermedia.com

EXECUTIVE EDITOR
Mike Senese
mike@makermedia.com

SENIOR EDITORS
Keith Hammond
khammond@makermedia.com
Caleb Kraft
caleb@makermedia.com

EDITOR
Laurie Barton

PRODUCTION MANAGER
Craig Couden

BOOKS EDITOR
Patrick Di Justo

CONTRIBUTING EDITORS
William Gurstelle
Charles Platt
Matt Stultz

CONTRIBUTING WRITERS
Evan Ackerman, Ayah Bdeir, Gareth Branwyn, Rich Cameron, Dr. Lindsay V. Clark, Larry Cotton, Sam Groveman, Joan Horvath, Shing Yin Khor, Bob Knetzger, Clare Mason, Meia Matsuda, Forrest M. Mims III, Michelle Muratori, Niq Oltman, Erin St. Blaine, Ulrich Schmerold, Micah Elizabeth Scott, Tucker Shannon, Jun Shéna, Andrew Stott, Eduardo Talbert, Sarah Vitak

DESIGN, PHOTOGRAPHY & VIDEO

ART DIRECTOR
Juliann Brown

PHOTO EDITOR
Hep Svadja

SENIOR VIDEO PRODUCER
Tyler Winegarner

LAB INTERN
Holden Johnson

MAKEZINE.COM

ENGINEERING MANAGER
Jazmine Livingston

WEB/PRODUCT DEVELOPMENT
Rio Roth-Barreiro
Maya Gorton
Pravisti Shrestha
Stephanie Stokes
Alicia Williams

CONTRIBUTING ARTISTS
Mark Madeo

ONLINE CONTRIBUTORS
Chuma Asuzu, Jennifer Blakeslee, Chiara Cecchini, Jon Christian, Leo DeLuca, DC Denison, Edwin Dertien, Rob Goodman, Liam Grace-Flood, Caroline Guenon, Cynthia Jones, Pinguino Kolb, Joel Leonard, Elizabeth Loose, Gina Lujan, Goli Mohammadi, Elisa M. Nicandro, Yukin Pang, Tasker Smith, Andrew Terranova, Zach Wendt, Jared Yates, Yan Zhang

PARTNERSHIPS & ADVERTISING
makermedia.com/contact-sales or partnerships@makezine.com

DIRECTOR OF PARTNERSHIPS & PROGRAMS
Katie D. Kunde

STRATEGIC PARTNERSHIPS
Cecily Benzon
Brigitte Mullin

DIRECTOR OF MEDIA OPERATIONS
Mara Lincoln

DIGITAL PRODUCT STRATEGY

SENIOR DIRECTOR, CONSUMER EXPERIENCE
Clair Whitmer

MAKER FAIRE

MANAGING DIRECTOR
Sabrina Merlo

MAKER SHARE

DIGITAL COMMUNITY PRODUCT MANAGER
Matthew A. Dalton

COMMERCE

PRODUCT MARKETING MANAGER
Ian Wang

OPERATIONS MANAGER
Rob Bullington

PUBLISHED BY
MAKER MEDIA, INC.
Dale Dougherty

Copyright © 2018 Maker Media, Inc. All rights reserved. Reproduction without permission is prohibited.
Printed in the USA by Schumann Printers, Inc.

Comments may be sent to:
editor@makezine.com

Visit us online:
makezine.com

Follow us:
🐦 @make @makerfaire @makershed
google.com/+make
makemagazine
makemagazine
makemagazine
twitch.tv/make
makemagazine

Manage your account online, including change of address:
makezine.com/account
866-289-8847 toll-free in U.S. and Canada
818-487-2037,
5 a.m.–5 p.m., PST
cs@readerservices.makezine.com

WELCOME

Why Shopping Mall Makerspaces Make Sense
BY MIKE SENESE,
executive editor of *Make:* magazine

In this issue, as we often do, we talk a lot about learning. A recent visit to Maker Faire Hong Kong offered a look at a learning venue that I hadn't previously considered, but one that should have been obvious.

Inside the glossy Olympian City shopping mall in Kowloon you'll find the same shops that populate just about every mall around the globe. At the top of the escalator on the third floor, however, stands a makerspace, containing 3D printers, electronics and soldering supplies, and even a well-filtered laser cutter. The space is called OC STEM Lab, and its purpose is to utilize the location — situated directly below soaring residential towers and alongside a popular subway stop — to offer hands-on learning to a community that is hyper-focused on increasing technology-based opportunities to its younger generation.

It's not the first makerspace in a mall — makers in Thailand and Malaysia have both leveraged commercial locations for their own, as did the D.C. TechShop. But its integration into the mall infrastructure makes it feel like an integral part of the experience, and makes it feel … just right.

This is a trend I'd love to see continue. Activity spaces offering technology tools, right where the people are. This one comes with the support of sponsors, and is run by the marketing team of Olympian City. They're proud of it, and very excited about the work that's been produced since its launch last fall.

CONTRIBUTORS

What's the most valuable maker skill you didn't learn in (traditional) school?

Clare Mason
Seattle, Washington
(Flashy Fashion)
I learned CAD through internet videos. The first thing I modeled was a fidget spinner. It was cool at the time, I swear. Now I model other stuff.

Evan Ackerman
Bethesda, Maryland
(Crash Course)
When I was in school, I had no idea how much fun it was to make things. I learned skills like coding and soldering later — the maker community is a fantastic teacher.

Meia Matsuda
Richmond, California
(1+2+3 Inner Tube Earrings)
Growing up in England, Japan, India, and the United States I didn't have traditional schooling. I'm currently in Richmond, California and adapting to unfamiliarities is my most valuable maker skill.

Tucker Shannon
Corvallis, Oregon
(Write on Time)
The most valuable skill I learned was how to Google search effectively. Even if you're doing something never done before, chances are someone has done something similar and posted a tutorial online.

Issue No. 65, Oct/Nov 2018. *Make:* (ISSN 1556-2336) is published bimonthly by Maker Media, Inc. in the months of January, March, May, July, September, and November. Maker Media is located at 1700 Montgomery Street, Suite 240, San Francisco, CA 94111. SUBSCRIPTIONS: Send all subscription requests to *Make:*, P.O. Box 17046, North Hollywood, CA 91615-9588 or subscribe online at makezine.com/offer or via phone at (866) 289-8847 (U.S. and Canada); all other countries call (818) 487-2037. Subscriptions are available for $34.99 for 1 year (6 issues) in the United States; in Canada: $39.99 USD; all other countries: $50.09 USD. Periodicals Postage Paid at San Francisco, CA, and at additional mailing offices. POSTMASTER: Send address changes to *Make:*, P.O. Box 17046, North Hollywood, CA 91615-9588. Canada Post Publications Mail Agreement Number 41129568. CANADA POSTMASTER: Send address changes to: Maker Media, PO Box 456, Niagara Falls, ON L2E 6V2

PRINTED WITH SOY INK

A Deep(er) Dive Into
Solar Power

Forrest M. Mims III

SOLAR POWER DRAIN

One thing Forrest missed in the article ["Small-Scale Solar Power," *Make:* Volume 61, page 72] is that by placing the solar panel behind the windshield like the photos show there is over a 50% drop in efficiency. Put it on the top of the car if you can. You need it perpendicular to the sun without any glass in between. —*Barry Klein, via the web*

Author Forrest M. Mims III responds:

Barry is correct about the absorption of windshields. On a day when the sun was 27° above the horizon, I placed a miniature 3V silicon solar cell array normal to the sun (perpendicular to its rays) in open air and behind the windshield and a side window of my Ford F-150 pickup. The current from the array behind the windshield and the side window was, respectively, 45.6% and 48.5% of the full sun current.

The spectral response of most silicon solar cells peaks in the near-infrared around 850–900nm. (Though some cells peak in the visible, where automotive glass is much more transparent.) An internet search of the spectral transmission of automotive glass provides similar results,

but some automotive glass provides slightly higher transmission in the near-IR.

Barry's suggestion about placing a panel on the roof of a vehicle will work best when the sun is high in the sky. The panel behind the windshield shown in the photo in my column was placed much more normal to the sun than if it had been mounted flat on the roof. This made up for some of the loss caused by the glass. Then there's the problem of theft of a rooftop panel when the vehicle is parked, and the need for a sturdy mount when the vehicle is moving.

Curious, I added ultraviolet radiation to my window transmission tests. A Solarmeter 6.5, which detects the erythemal UV wavelengths that cause reddening of the skin (*see "DIY Sunburn Sensors" on page 62 of this issue*), was pointed directly toward the sun through a windshield and side window. In both cases, the meter indicated a UV Index of 0.0. A Microtops II photometer that measures UVB at 300, 305, and 312nm indicated no signal through the windshield at all these wavelengths; 16% transmission through the side window at 300nm and no transmission at 305 and 312nm.

Space - Robot - Code
@wonrobot   Follow ∨

A few days ago, Arushi was reading the interview of #battlebots wiz @lisawinterx in the @make magazine. Today @Fitc #futureworldtoday she got to meet Liz and give a demo of her #KiloNova bot. More bot info at hotpoprobot.com

2:52 PM - 9 Jun 2018

3 Retweets 17 Likes

♡   ⊔ 3   ♡ 17

HAVE SOMETHING TO SAY?
We want to hear!
Send us your stories, photos, gripes, and successes to
editor@makezine.com

MADE ON EARTH

Backyard builds from around the globe

RETRO RACER

INSTAGRAM.COM/JIMMY.BUILT

Jim Belosic's dad was a "car guy." His dad's dad was a car guy. And he grew up playing with and disassembling every type of R/C vehicle he could get his hands on. So it is no surprise that he had a strong connection to his first car, a 1981 Honda Accord, despite the fact that it only lasted in his possession for about 6 months.

Since then, Belosic has continuously worked on cars. But in the past couple of years he began to grow bored. "You can only swap engines or whatever a few times before it's not that exciting anymore," he says. "I started to explore what else was out there besides V8s and internal combustion engines." This led him to shift gears last year and build a steam-powered engine completely from scratch. He enjoyed the chance to learn something new. "It really satisfied the maker in me," he says.

Out of nostalgia, Belosic bought another 1981 Accord in 2014, and decided late last year that he wanted to use it for a new project. But he wasn't quite sure what. "My son is now 8 years old and loves to wrench with me, so I started thinking about what kind of cars we'd be working on together in the future," he says. "I believe that electric cars are the future, and I started to get nervous thinking about trying to help my son work on his weird electric car when he's 16." And then it clicked. He would convert the Accord to electric and gain some new skills in the process.

He decided on a Tesla Model S drivetrain and a battery from a Chevy Volt. The Volt battery is more "power dense" than the more expensive Tesla battery, and it was the perfect size and shape to fit inside the engine bay.

All in, it took Belosic and his friend Mike Mathews about 2 months to get the first running version. Mathews is responsible for the RasPi dashboard with awesome retro-gaming graphics. They have been working on reiterating and refining the project ever since. The finished "Teslonda" weighs in at just under 2,500 lbs — about half that of a Model S — and has been clocked reaching 60mph in 2.48 seconds. You might just call it ludicrous. —*Sarah Vitak*

Keiron Berndt

Know a project that would be perfect for Made on Earth? Let us know: *makezine.com/contribute*

makezine.com 7

FLAWLESS FUSION

SCOTTSLAGERMAN.COM

There is something inexplicably captivating about **Scott Slagerman**'s Wood and Glass series. The fluidity of the two mediums is magical and is clearly due in no small part to Slagerman's artistic brilliance. He is a classically trained glassblower who has always been drawn to the large wooden beams used in construction and wanted to find a way to bring them into contact with glass.

According to Slagerman a lot of the time spent creating each piece is just holding the materials next to each other, drawing, and visualizing the final product. And that says a lot, because the creation of each piece has many time-consuming steps. Once he has a design in mind, he will draw the shape on the wood and carve it out, usually trying to run along the grain. From there he will prepare the wood and blow the glass shape directly inside of the wood. When the glass reaches precisely the right temperature it is removed and put in a kiln to cool. The glass will have a final fitting where it is cut with diamond tools before ultimately being placed in the wood. And finally, the wood will be sanded and oiled.

Slagerman has worked with many different types of wood for the project. He began with construction grade lumber and monkey pod (a tropical hardwood) that was imported from Fiji. Since then he has begun collaborating with his friend Jim Fishman and has experimented with many wood sources and types. Slagerman has other projects that he is working on concurrently, including some incredible wood and glass joinery. But for this series he says, "I'm just still advancing with making new pieces and seeing where I can push the limits." —*Sarah Vitak*

Scott Slagerman

DIGITAL MEETS ANALOG LEXOPTICAL.COM

Photographers often try to pair vintage, manually controlled lenses with state-of-the-art digital camera bodies in attempts to get classic-looking results while using the most advanced features. Texas-based **Alex Gee** is taking that concept and flipping it, creating a film-based camera body for his digital-only Sony E-mount lenses.

The project began last year, utilizing an Arduino Pro Mini and a custom PCB to interface with the lens aperture electronics and shutter speed mechanism sourced from a Sony A7 camera. He's produced a simple 4-button control panel that lets him adjust the settings, displaying them on a small screen on the rear of the camera, and trigger the shutter.

Initial fabrication of the body itself occurred on Gee's Monoprice 3D printer; once refined, he combined laser-sintered body parts with metal components that he cast himself. The result is a stylishly retro camera body that's slender and simple — except for the fancy lenses that twist onto the front.

Gee is also keeping the project open source, sharing files on GitHub. "What we are aiming for is a solid jumping off point from which people can iterate on things without having to worry about making 10 different prototypes just to get the film canisters to fit snuggly," he says, explaining his hopes that others will contribute to the development. "Ideally we want this to remain easy enough to build that it can be a high school electronics lab project." —*Mike Senese*

Danielle Baskin

HOT WINGS DRONESWEATERS.COM

A sweater on a hobby drone? A joke, right? It was when self-described professional prankster **Danielle Baskin** first thought of it. She was asked by a friend if she was doing anything for Fashion Week, and knowing that drones were trending tech at the time, the absurd idea of drone sweaters popped to mind.

Baskin put up a website as part of her spoof, offering drone sweaters for sale. And amazingly, she started getting orders. OK, not everyone rushed to buy a sweater for their drone. To date, Danielle has only sold 5 (for $89 each). But that hasn't stopped her from continuing to advance the cause of drone sartorial. She is also now

working on drone coats, drone skirts, and even a drone suit for the upwardly mobile executive drone.

Flush with the positive response she got from her drone sweater caper, Baskin wondered how digital daters might respond to a clothed drone on a dating site. She posted a profile on Tinder and it wasn't long before her drone started attracting suitors. Baskin actually sent her drone on dates and hid where she could see the drone and the date, listen-in, and respond, using a Bluetooth speaker and microphone hidden underneath the drone. At least one dater asked for a second date. But the drone might be too busy to date, soon. Baskin

thinks it's time her quadcopter got a job and is thinking of sending it out on job interviews.

Besides being a professional prankster, Baskin also describes herself as a "situation designer." She has been behind a number of other whimsical and inspiring projects such as Last Chance Tours, VR tours of buildings about to be razed, Goodbye Domains, a graveyard for expired URLs, and Custom Avocados, where she laser-prints logos and messages on avocados. Her latest endeavor is Your Boss, an app that allows freelancers to support, coach, and cheerlead each other.
—*Gareth Branwyn*

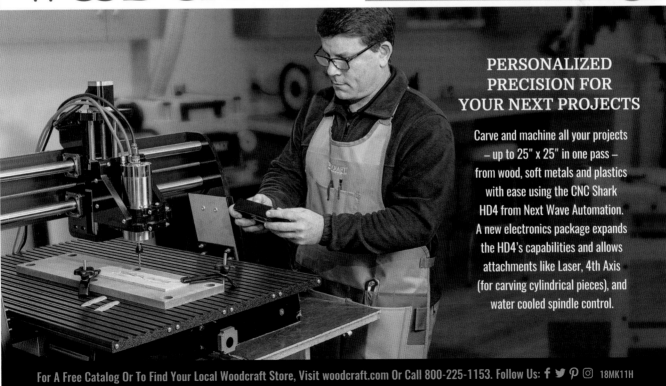

Founder and CEO of littleBits, **AYAH BDEIR** is an engineer and interactive artist.

littleBits Goes Big

desire for accessible, creative electronics helped bring her maker building blocks into the mainstream

Written by Ayah Bdeir

I GUESS I WAS ALWAYS A MAKER WITHOUT KNOWING THAT IT WAS A THING. I first learned about the maker movement in the first class I took at MIT's Media Lab in 2004. It was called "How to Make Almost Anything" and it was one of the most competitive classes to get into at the Media Lab. Taught by Neil Gershenfeld, the current director of The Center for Bits and Atoms at MIT and the father of the FAB movement, and to many, the maker movement, the class was featured in the inaugural issue of *Make:* magazine. And just like that, I was witness to the birth of a movement that would change my life.

I identified with the "maker" label right away because I had been *creating* my whole life. I used to break open VCR players and gravitate to all things construction when I was a kid. My educational background was officially in engineering but I was always interested in how engineering could relate to other fields: social, political, artistic. I suddenly realized that there was a whole community of people with my interests — and I became addicted to it.

Before the Bits

I grew up in a loving home with three sisters where my parents were incredibly invested in developing our curiosity. When we were young, we were exposed to other languages and countries; we traveled extensively and learned often about the world around us.

We lived in Lebanon, a country with enormous potential but hindered by its own sectarianism, lack of cohesion, and short-sightedness. Lebanon doesn't have any natural resources, like oil, so its people are very entrepreneurial. They come from generations of traders; often, they are comfortable speaking multiple languages, traveling, making connections with other people, etc.

I think that it's in the Lebanese nature to be an entrepreneur. We are resilient. We love life. Despite experiencing a lot of hardships and strife, we rise up and find joy in even the most difficult of circumstances.

After graduate school, I landed a job as a financial-software consultant for a tech company. The work was unfulfilling and intangible, so I immediately began looking for other ways I could pursue my creative interests. I wanted to make things!

I secured a fellowship from Eyebeam, one of the pioneering labs of art and technology, with the goal of getting back into making. It occurred to me that there were probably a lot of people in my position: stuck in roles that didn't allow them to use their creativity. While I had the technical background to push past that, others did not.

That's when I got the idea to make electronic components that could be used by anybody — regardless of age, gender, or technological background. Not only would I be able to make things again, but I would be helping others to make things as well.

A Small Experiment

The very first version of littleBits was created while working with Jeff Hoefs, at Smart Design, and it was never meant to be a product. It was just a small experiment to get designers to prototype and invent with electronics before we moved on to more "traditional" interactive design work.

I debuted a prototype of littleBits at Maker Faire 2009 with interactive designer and founder of Chibitronics, Jie Qi. She was an intern with me at Eyebeam. All of the Bits we brought to our booth were handmade; they kept failing and we kept having to jiggle them to make them work. To be honest, we felt like frauds at that first Maker Faire. We felt like we were showing off a concept that didn't work. And that propelled us to work harder to get the prototypes improved.

But at the same time Maker Faire was an incredible experience because I needed the feedback to iterate. Kids were lining up at our booth to get a glimpse of our Bits. We even set up a wall called "Submit your Bit" which invited people to suggest the Bits we should create next, from a light sensor to a buzzer or an LED matrix to a "fart sensor" or even a "penguin finder." It was so exciting to think of the possibilities.

Mark Madeo

All In

By 2011, littleBits was ready for prime time. *The New York Times Magazine* wrote a piece that was published in May, and by June I had quit my job and devoted my time and energy to littleBits full-time — I got my first prototype off the factory floor! The bitSnap (magnetic connector) was made in China and the first Bits were made by SparkFun.

I thought it would be smooth sailing from there. The factory prototype was ready, now all I needed was to place some orders and ship them. Piece of cake, right? Boy was I wrong! There is so much to making and shipping a product beyond the first factory prototype. I always say the MVP of a hardware company is a hardware company. You can't be a couple people in a garage writing code like in software, you have to have it all figured out: quality control, customer service, warehousing, shipping, taxes, return labels. When people pay for something their expectations are very high from the get-go. And you don't get to "send updates" by internet.

littleBits, the company, officially launched in September 2011. We did our first product sale at Maker Faire New York. It was just myself and Paul Rothman, creative technologist and very first OG Bitster.

It was a proud moment; we were launching a new category: electronic building blocks. And initially, it was relatively easy to raise money. Joi Ito, current director of the MIT Media Lab, saw littleBits in a demo and emailed me about investing. From there, I was able to quickly get some other investors on board for an initial round of $850K. The vision was always the same: inspiring people to be creative with electronics. Investors were as enamored as I was with littleBits' potential.

In October 2011, we were acquired by MoMA for its collection. Our first edition Bits now had a permanent home snuggled in between a Picasso and the Post-It — not to mention being included in a contemporary interactive exhibit, *Talk To Me*.

Startup Challenges

I think any entrepreneur goes through the impostor syndrome multiple times in their journey. You are learning so many things for the first time. How to make a product, how to scale, how to hire, who to hire, how to sell investors, how to sell individuals. And

"The vision was always the same: inspiring people to be creative with electronics."

INVENT HERE.

littleBits

"All of the Bits we brought to our (Maker Faire) booth were handmade; they kept failing and we kept having to jiggle them to make them work. To be honest, we felt like frauds at that first Maker Faire."

1. littleBits' NYC-based pop-up shop: The Invention Lab (2015).
2. An example of the simplest littleBits circuit: power, input, and output.
3. Ayah Bdeir assembling an early version of littleBits at Eyebeam (2009).
4. Kids in class working with the littleBits Code Kit (2017).
5. Inside The Invention Lab.

you are constantly thinking: everyone has all their stuff together except for me. Why is this so hard?

And the truth is, it just is. It's hard for everybody. Some people are just better at faking it than others. So you have to get your validation where you can. One of the biggest sources of validation for me has been the community. A few months after littleBits launched, I got a Google Alert for a YouTube video showcasing a project a boy invented with his dad. I watched it, then decided to search YouTube for the word "littleBits." Suddenly, I found all sorts of projects that people had made — from South Africa to Singapore, Mexico to Canada! It was such an exciting moment for me to realize that littleBits is a universal product that was already making an impact on a global scale.

Even today, we still get a lot of fan letters from all over the world — from parents who have seen their kids suddenly take an interest in STEAM, to kids who have discovered that they are more creative than they think, to people who finally feel like

technology is accessible to them. Teachers also report that kids who are struggling in class are often the ones who become the most engaged with littleBits.

There are decades of research that show that learning through play is extremely effective. It's energizing to realize that littleBits is playing even a small part in helping students to connect with STEAM and find their "moment" — when they snap together a circuit for the first time and realize that they can create anything.

When I see children creating inventions that will make a difference in someone else's life — anything from a helmet to help the visually impaired to a bicycle that could prevent accidents — it's truly inspiring.

Company and Community
For a full 18 months after littleBits launched, I still called myself the company's lead engineer. One day, I realized I was doing very little engineering, so I accepted the title of CEO. Now, I am excited about how far other people have taken littleBits

— beyond what I would have ever imagined. There's so much to be learned from other perspectives. Today, I am able to embrace that and focus on other things.

Community building has been a natural progression. I am passionate about littleBits' mission, so I take any opportunity I can to connect with like-minded communities — maker community, STEM community, girls empowerment community, entrepreneurial community — to create synergy. I've learned so much from being a part of these communities. Namely, the best way to be a good community builder is to be a good community contributor.

littleBits has experienced many iterations on its path to becoming the company it is today. At different times in our history we've celebrated different successes and faced different challenges. We've had multiple near-death experiences, from manufacturing disasters that propelled a group of us to take overnight flights to China within a couple of hours, to retailers we had big partnerships with shutting down from

Mark Madeo, littleBits, Jun Shéna

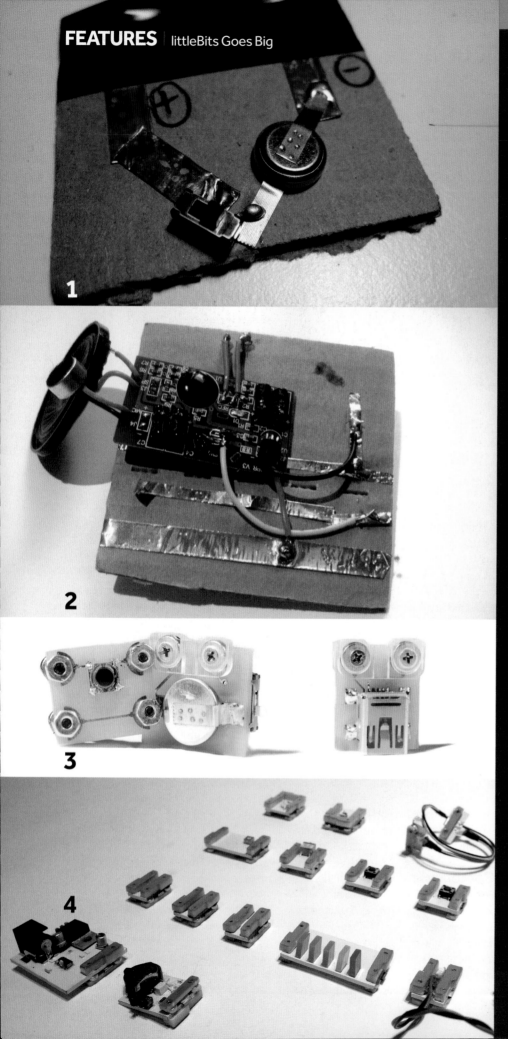

1

2

3

4

> "It's energizing to realize that littleBits is playing even a small part in helping students to connect with STEAM and find their "moment" — when they snap together a circuit for the first time and realize that they can create anything."

AYAH'S TIPS TO GET MORE GIRLS INTO STEAM:

1. Start early. The National Girls Collaborative Project reports that girls score almost identically to their male classmates on standardized tests through high school. Yet, boys demonstrate twice as much interest in STEAM careers as girls as early as the eighth grade. We believe 8-year-old girls are the tipping point; and the earlier you capture girls imagination the more impact you can have.

2. Make a gender-neutral product. There are more than enough gendered products. They don't encourage kids to play together, they reinforce stereotypes and they don't expose kids to diverse interests. Creativity is not gendered, curiosity is not gendered, STEAM should not be gendered.

3. Empower girls to inspire other girls. Create content and marketing materials that showcase empowered women using your product — helping girls to envision themselves doing the same. Make a conscious effort to show girls in leading, not secondary, roles. Show them that building and inventing is fun and exciting, and that there's a place for them in that world.

1. and 2. First-generation cardboard prototypes.
3. Second-gen prototypes get more refined.
4. Fourth generation shows the signature design.
5. Ayah Bdeir working with teammates Ryan Mather, Aya Hamdan, and Sarah Page.
6. Inside the littleBits office: Emily Tuteur, Evan Spiller, Carrena Nunez at work.

littleBits., Mark Madeo

one day to the next and all of our products going on clearance. To that end, there is always something different, a new frontier to cross, a new challenge to overcome, a new lesson to learn.

But there are also lots of things that have been consistent. For example, ever since the beginning, we have always been laser-focused on the design of the Bits. It was important to me that they were gender-neutral. From the color of our circuit boards to our packaging, each of our products has been deliberately designed to appeal to both girls and boys. This level of accessibility helps everyone to unleash creativity and to instill a love of STEAM through inventing.

As a result, we've been able to attract an unprecedented number of girls into STEAM — and develop a beautiful product. Today, and consistently over the past four years, 35%–40% of our user base is girls; that number is four times the industry average!

New Partnerships

Today, the littleBits platform includes more than 10 kits and over 70 interoperable Bits; our team is over 100 people strong. To date we've sold millions of products to inventors in more than 150 countries around the world; we have more than 300 littleBits Inventor Clubs from Sao Paulo to San Francisco to Bangkok.

In the summer of 2016, littleBits was invited to participate in the Disney Accelerator program, which seeks to identify and support "innovative consumer media and entertainment product ideas." It has proved extraordinarily successful in incubating both early stage companies and later-stage businesses like littleBits.

We worked closely with Disney to create our first licensed product, the littleBits Droid Inventor Kit. This is not a *Star Wars* replica toy; it's about the building and making of inventions that promote STEAM through iterative education. For the first time ever, Disney allowed a licensee to "go off-book" and create a character-based toy that could be customized. From the core product to the pictures on the box, Disney worked with us to embrace the value that littleBits could bring to its portfolio of licensed products.

Star Wars inspires its fans in different ways. Tinkering, scavenging, and invention are all timeless themes in the franchise. Together with Disney and Lucasfilm, we saw a huge opportunity to mimic the science and

technology and invention ethos in a hands-on product for current and future fans of the franchise. The Disney Accelerator program has changed the course of littleBits and showcases how we could reach outside the early adopter crowd and into pop culture and a mass audience. The product won The Toy Association's Creative Toy of the Year award this year, in addition to being among Amazon's top 10 holiday toys, the #1 *Star Wars* toy, and the #1 toy over $50 in Q4 2017. It has also been named to several of the toy industry's most prestigious "hot toy" lists. We've just launched a second licensed product in partnership with Marvel Entertainment, the Avengers Hero Inventor Kit, which started shipping this month.

Tools to Grow

Over the past decade, we have seen huge changes in the meaning and the role of education. littleBits has positioned itself at the forefront of a new generation of companies — driven by tech, design, and the maker movement — who are redefining the future of learning through play. To that end, we've recently made a very big move: we made our first acquisition, DIY.org. This is one of the largest safe online communities made for kids to share content, discover new passions, level up their skills, and meet fearless geeks just like them. Together we

are reinventing kids education: at school, at home, and everywhere in between.

And most importantly we stand for our community. Today, littleBits has been used in millions of inventions around the world. We have a huge community of inventors who are dedicated to our brand; we are authentic; and importantly, we don't stop at incorporating science, technology, engineering, and math into our product — we also make sure to incorporate art, music, and creativity into everything we do.

One big thing we learned, for instance, is we have the ability to reach kids where they already have interest — helping their creativity come to life. 10-year-old Sky from Southern California is into medicine, so she made a Medicine Machine that dispenses water, medicine, and tracks the amount of medicine taken. Enxhi, an 18-year-old from Gjakova, Kosovo, loves fashion; she designed a wearable electronic skirt that lights up using littleBits. An 11-year-old inventor named Vedant, from Danvers, Massachusetts, said about littleBits: "My imagination can be real now." He had devised a way to control his Lego cars with a remote control he built using littleBits.

I'm proud that the littleBits team has been able to make such a profound difference in the lives — and inventions — of makers around the world. ⊘

Newton's 3D Printer

Rethinking calculus curriculum to make it more accessible using physical models

Written by Joan Horvath and Rich Cameron

JOAN HORVATH is an MIT alumna, recovering rocket scientist, and educator, and **RICH "WHOSAWHATSIS" CAMERON** is an open-source 3D printer hacker who designed the RepRap Wallace and Bukito printers. They are the co-founders of Nonscriptum LLC (nonscriptum.com), and are currently writing their 7th and 8th books together.

EDUCATORS HAVE A GOLDILOCKS PROBLEM WHEN THEY TRY TO INCORPORATE 3D PRINTING IN THEIR CLASSROOMS. Students don't learn a lot from downloading a model and printing it. We've also seen free models capture a concept in a way that is misleading at best.

At the other extreme, asking students to create a model with a CAD program requires teaching far too much about the mechanics of model making for many teachers' taste. It also takes deep knowledge of the science and math to do a good job.

We realized we could start creating "just right" models by melding Joan's math and science background with Rich's knowledge of 3D printing and geometric problem-solving. Our models are not just "download and print," but do not require a student to start from absolute zero, either. We design models that can be altered based on the science, not just scaled to fit a printer. We also write about the science presented by the model as well as the assumptions and limitations.

We've now written two books that are collections of these model-based science projects, with models written in OpenSCAD. We try to create our models so that they will print without support on a poorly tuned, abused 3D printer, like the ones found in many schools. In every chapter, we included a "Learning Like a Maker" section — all the mistakes we made trying to create the model, so that anyone altering it can know what we tried that didn't work.

Kepler's Reprise

It sounds obvious when you say it, but our first key takeaway was that 3D prints could be used to create true three-dimensional graphs, which could be a mix of physical and abstract dimensions. One of our favorites is what we call "Kepler's Laws in Plastic."

In the early 1600s, Johannes Kepler realized that planets travel around the sun in ellipses, with the sun at one focus. His data showed that planets went faster near the sun, and slow down as they move out into the dark farther away. More specifically, his second law says that a line from a planet to the sun will sweep out equal areas in equal times as the planet goes around the sun. We designed the models shown in Figures A and B to visualize what the variation of speed around the orbit might look like.

The bases of these models are the orbits of bodies around the sun, and the height is how fast that body is going at that particular point in the orbit. The long, skinny orbit in Figure A is the orbit of Halley's Comet — you can see the speed ramp up near one end, where the sun is, and drop way down out in the far darkness away from the sun. The set of three models in Figure B is the orbits of Earth, Venus, and Mercury. Mercury's orbit is a little elliptical, and so it goes a bit faster at one end than the other.

We've also created models on topics ranging from plant growth (Figure C) to wings, molecules, gravitational waves, sand dunes, logic circuits, and more. We were struck by how much we learned developing the models, manipulating them, adjusting, and thinking about what to show and what to hide.

The Ultimate Tactile Learner

Teachers of visually impaired and blind students were interested in our models early on. Considering how someone blind would use a model naturally eliminates some of the really bad applications of 3D printing we've seen, like printing raised-line perspective drawings of 3D geometrical objects, instead of just printing the cone or cube in the first place (Figure D). It also sensitized us to the storytelling aspect of a model. How is it meant to be handled? Is there a starting point?

We had a workshop recently at a scientific conference about making models like this, with one participant who happened to be blind. The presenters brought along 3D printed models that made their key points and gave them to our blind colleague as they discussed them. It made it much more possible for her to participate than just trying to imagine PowerPoint slides. Beyond that, though, many sighted participants came up afterwards to explore and play with the models, and it was clear there was a lot more understanding and engagement than typical from everyone.

Next Up: Channeling Newton

In the course of creating some of our models, we ended up deriving calculus concepts from first principles. Rich began making exploratory models to help think about these, and we realized we were creating a new way to teach an end-to-end, alternative, hands-on calculus course. For inspiration, we returned to the first calculus book, Isaac Newton's 1687 *Principia Mathematica*. To our surprise, we discovered there was almost no algebra — but many geometrical diagrams.

Using 3D printing as a lever to change math and science teaching requires significant reordering of concepts. For our restart, we are going as far back as Newton. We'll build intuition with geometry-focused models and physics analogies rather than pages of algebra manipulation and rote memorization of formulas. We call this project "Hacker Calculus," with the tagline, "What if Isaac Newton had owned a 3D printer?" We're currently writing a book for MIT Press, and are putting our models out open source to create a community for change. We think Isaac, quite the maker himself, would have approved. ◓

tiero - Adobe Stock, matiasdelcarmine - Adobe Stock, Figures A,B,C from *3D Printed Science Projects*, Apress; all photos Rich Cameron

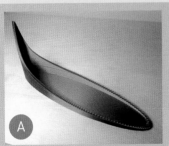
A

B

C

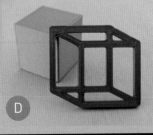
D

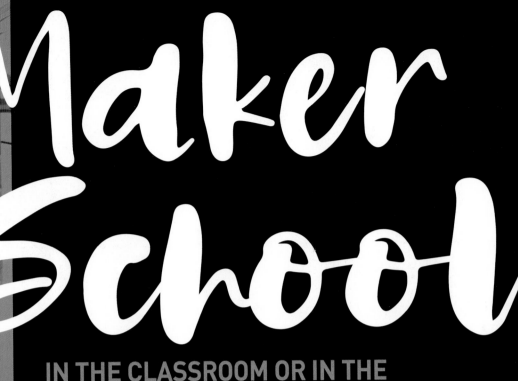

Rawpixel.com - Adobe Stock

Maker School

Written by Mike Senese

IN THE CLASSROOM OR IN THE WORKSHOP, THERE'S GREAT VALUE TO LEARNING FROM OUR PROJECTS

The resurgence of making as an activity — and as a definition of self — has matured into an active component of our global society. It's touching all areas, from the hobbyist level to large industry. Perhaps most prominently, making can now be found in learning and education; schools on all continents are adding Maker Ed curriculum to their programs along with their own makerspaces, giving students immediate access to new technology tools, and the guidance to incorporate them into their own studies and work. But as this happens, we must not lose connection to our foundation.

In her piece "Making Is Where Work and Play Are in Perfect Harmony," [makezine.com/go/measurably-useful-work] Dutch artist/author Astrid Poot petitions for playfulness as making transforms into a serious element of our academic and work environments. She offers guidelines, including the following:

Edu-entrepreneurs, if you want kids to make, then respect the beauty of it. Do not stifle it with processes and methods and measurability. **Trust the makers, no matter how young they are.** *Do not focus purely on short term utility, but focus on growth of the students (long term). Give teachers the inspiration to do the work* **in their own ways.** *Then it will become useful on its own. And many of the 21st century skills have already been a big part of education anyway. Embrace that.*

We agree.

Learning also isn't confined to the classroom, nor is it restricted by age. That's an important part of the maker mindset. We learn best by doing, by using our hands, and by following our passion for creating something — be it useful, delightful, or both — we acquire new skills. These can be transferred to the next project, making it better, faster, cheaper. And they can be transferred from our personal endeavors to our professional careers. From coding to design to fabrication, there is great value in making.

We encourage you to embrace these interests, to become a master of your passions. Take time to understand the principles of your project, and how to incorporate them elsewhere. Become an expert, and share your expertise with others. Practice this and good things will come to you, to your community, and to the maker movement as a whole.

Crash Course

BUILD YOUR SKILLS FOR MAKER SUCCESS

Written by Evan Ackerman, Caleb Kraft, and Hep Svadja

Learning a new vocation in today's economy is easier than ever, with online communities and resources to teach us and get up to speed on the new technologies, and help us expand our reach. There are a lot of ways to get started with a maker hobby, but the task of becoming an expert, and even going pro, may seem overwhelming. If you break things down into steps, however, the path becomes a bit clearer.

To that end, we've asked authorities in different areas of making — from classic crafting to robotics and cosplay — about how they navigated their fields, and to share some of the opportunities that await a maker who's looking to advance with their craft.
— Caleb Kraft

EVAN ACKERMAN is a freelance science writer based in Washington, D.C. Since 2007, he has written over 6,500 articles on robotics and emerging technology.

21st Century Robotics

If you ask five roboticists what they think a robot is, you're likely to get 10 different answers. The most generally accepted definition for a robot is probably something like, "a mechanical system that can sense its environment, and then based on what it senses, autonomously take actions that affect its environment."

Any robot, no matter how basic or how complex, is a combination of software and hardware. "One of the things that makes robots special is the system nature of it," says Colin Angle, co-founder and CEO of iRobot. "It's not just about software, it's not just about sensors, it's the whole system." This can make robotics a little bit intimidating, because it seems like you'd need to learn mechanical engineering and software engineering at the same time to do anything useful.

Fortunately, much of what's interesting and creative in robotics today is happening in programming, and the gear isn't nearly as hard as it used to be. "Right now the biggest advancements in science are on the software side, much of which were enabled by breakthroughs on the hardware side years before," says roboticist Carol Reiley. "Many industries are transitioning into a software-driven/AI-powered future. There's a lot of opportunity for growth on software and understanding data in the physical world around us."

LEARN TO CODE

If you're already comfortable with writing code of any sort, that's fantastic, because writing code for robots is very similar to writing code for anything else. The specific languages to get experience in depends on what aspect of robotics you want to work on, says Brian Gerkey, CEO of Open Robotics. "If you're developing the autonomy software on the robot, then C++ and Python are the most common languages. But robots operating in the real world also need interfaces that are suitable for non-engineers, which are usually based on web or mobile technology, so JavaScript, Java, and Swift are good to know."

If you're not comfortable with writing code of any sort, that's also fantastic, because robotics is the most exciting way to learn. There are many ways to start without having to worry about writing code at all, using visual programming languages where you drag and drop interactive blocks to build a program. The most well-known system like this is probably Scratch, which was developed by the MIT Media Lab. Blockly, from Google, has the added benefit of showing you a real-time translation into a variety of different coding languages, including Python and JavaScript. Scratch, Blockly, and visual programming languages like them are the interface of choice for most introductory robotics kits, and they'll

Hep Svadja

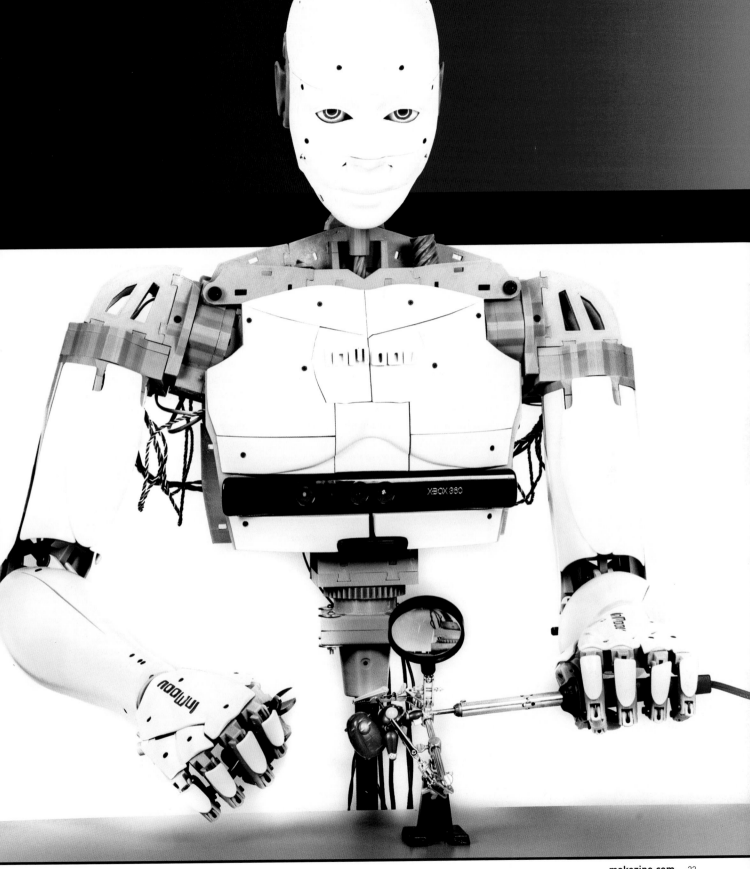

1

1. Ozobot Evo
2. Finch
3. Root
4. Lego Mindstorms EV3
5. Vex IQ Super Kit
6. iRobot Create 2
7. Misty II
8. TurtleBot

2

3

software as well as hardware, and to really appreciate how to make that work effectively, you'll need some experience with both. "It's important that as you get started in robotics, you get exposure to everything working together," says Angle. "Get your hands dirty and start building things, or go find something, take it apart, and try to understand it well enough to put it back together again. That type of system understanding needs to be present, and then you can decide what area of specialization most speaks to you."

GETTING STARTED

The most complicated robots are made of lots of sensors, actuators, and autonomous logic. Simple robots are made of the same things, but just a few of them, and the simplest robot to make just has one of each. An easy way to get started is with an Arduino starter kit, as long as you find one that includes both a sensor and an actuator. Following tutorials that, for instance, take you through how to control a servo motor and how to read a temperature sensor give you the basis for a robot that autonomously turns on a fan when it gets warm.

However, people like things that move, and for beginners to both robotics and programming, we recommend you start with a pre-built platform on which you can practice your coding skills. You'll want one that includes several different levels of coding interfaces, allowing you to go from simple to complex as you get comfortable.

Ozobot Evo is particularly good for younger beginners, since you can "program" it by drawing patterns of colors on paper that the robot reads as it drives over them, without needing a software interface at all. From there, you can move on to a Blockly-based visual programming language. **Finch**, an educational robot developed at Carnegie Mellon University, is a bit more versatile, and includes light sensors, temperature sensors, and basic obstacle detection. Finch starts with a Scratch-based visual programming language called Snap!, modified to make it accessible to children in early elementary school. The Snap! interface gradually gets

can wander across wall-mounted dry erase boards (sticking with magnets), and comes with an impressive array of over 50 sensors, actuators, and interactive elements like touch-sensing surfaces and LEDs. Root starts out with a visual programming interface, which later shifts into text-based coding in Python, JavaScript, and Swift.

If you'd prefer more versatility and options in hardware along with a hands-on building process, the **Lego Mindstorms EV3** and the **Vex IQ Super Kit** are good choices. Both can be programmed in Scratch-like graphical environments, with options to continue in LabVIEW for Lego, and C (through ROBOTC) for both Lego and Vex. Lego and Vex also integrate with worldwide robotics competitions, offering a good way to further challenge yourself while meeting other people interested in robotics.

LEVELING UP

In order to do most of the really, *really* cool stuff in robotics, you'll need access to more sophisticated hardware, including better sensors, better actuators, and better computers. As you get more confident with writing code, you'll likely find that educational robots and most robotics kits simply don't have the hardware necessary to do everything you want to do. In the past, getting beyond this point would mean building your own robot, or using a starting point like **iRobot's Create 2** mobile base and

4

5

6

7

8

adding your own hardware on top. This is still a good option if you want to experiment with hardware.

Over the last several years, a new class of robots has been introduced, designed for students and experienced hobbyists, that combines powerful hardware with accessible software. These robots are probably not for beginners, but if you want to focus on writing code, they'll give you the most options. **Misty II** is specifically designed for people who have some programming experience but not necessarily any robotics experience. It is equipped with various sensors, and an expansion port on the robot's back plugs directly into an Arduino or Raspberry Pi, making it easy to mix in your own hardware accessories. **TurtleBot** is a development platform designed around the Robot Operating System (ROS). ROS was designed to be general purpose software (built on C++ and Python) that can power just about any kind of robot, relying on experts to collaborate and extend a robustly supported open source core. It's been remarkably successful, and TurtleBot is intended to be an affordable way to learn how to program with ROS. As such, TurtleBots are often used in robotics courses, and they're capable and expandable enough that you frequently see them as early prototypes

for robots that were later turned into commercial products.

WHAT'S NEXT?

There's plenty to do with robotics as a hobby on your own, but if you're looking for ways to take things further, finding a community (like a local robotics club) can provide both inspiration and support. If there aren't enough folks near you with an interest in robotics, online communities are an option as well, and can be especially helpful if you're stuck on something specific.

There are even more options for students. Vex and Lego (through the FIRST organization) run kit-based robotics competitions for teams starting as young as six years old. For high school students, competitions can get very serious, with thousands of teams from all over the world building and programming complex robots to solve challenges that are different every year.

Both FIRST and Vex offer a good combination of practical hardware and software experience, but there's more value to joining a competitive robotics team than just building a robot — being part of a group working together on a big project provides its own challenges and rewards. "Get hands-on experience [and] work on a team," says Marc Raibert, founder and CEO of

Boston Dynamics. "The ability to work with others is just as important as your technical skills." Angle agrees: "The first quality [we look for] is being able to work effectively in a team. Robotics is hard, it's multidisciplinary, and no one person is going to build the type of robot we build at iRobot."

Beyond high school, many colleges and universities are developing dedicated robotics programs, and a few schools even offer degrees in robotics specifically (as opposed to mechanical engineering or software engineering). Competitions continue to drive robotics education as well, from RoboCup's autonomous soccer leagues and challenges for home and rescue robots to competitions sponsored by companies like DJI and Amazon, looking for clever new technologies as well as new talent.

The best advice for those looking to extend their robotics passion into a career is to "start now," says Gerkey. "The demand in industry for people with solid robotics skills is growing fast, but so is the supply. We're seeing students who have been writing code and building robots for several years before even finishing high school. That's your competition, so get prepared." And if you want a job working on robotics at a place like Boston Dynamics? Simple, says Raibert: "Do what you love doing." — *Evan Ackerman*

Three Types of Crafts

Traditional maker activities are as enjoyable as ever, with ample opportunities for side gigs and more. Crafting is a diverse field; here are ideas on getting started with three distinct types — woodworking, cooking, and knitting.

WOODWORKING

With the internet seeding new interest in woodworking, the activity is seeing a resurgence as both a hobby and a career. The amount of educational and inspirational videos online is skyrocketing. While historically someone pursuing this field as their profession would find somewhere to apprentice (and some still do — see "Should You Apprentice?" on page 28), many are picking up the basics from a teacher or a few online videos, then honing their skills on their own.

"YouTube and Instagram are great places to learn or see different techniques, as well as sites like The Wood Whisperer," says Chris Brigham, who is a professional woodworker at Knife & Saw (his personal company) and FineRoot (where he's a partner). "I am constantly looking at other people's work and seeing how they do something differently than I would."

Get Dusty

If you want to get good with wood, you're going to have to force yourself to try new things. You'll make some ugly furniture. Your miter joints won't be perfectly matched for some time, and you will get frustrated. "I always think the best way to get started on something is just to jump in and try it," Brigham says. "Most likely it will be frustrating and what you make will be garbage but you will learn from it."

A great way to keep yourself going is to enter online contests. Podcasts, tool companies, and Instructables hold them frequently. Though you may not win awards (but maybe you will!) you will find that you push yourself to hone your craft, and often pick up a new trick or two.

Going Further

You'll eventually want to start marketing yourself. Brigham found early success with one of his projects getting a lot of attention online. "Some friends that I had shown The Bike Shelf to submitted it to influential design sites like Hypebeast, Selectism, Cool Hunting, Apartment Therapy, etc. and it kind of went crazy from there," he says. "Because of that press, I have been able to get jobs doing custom work."

As you grow you'll likely find that you have an aptitude for a particular item's construction. Either your style stands out or you're just really good at making something. Don't be fooled into thinking it has to be something fancy or rare. There are cutting boards on eBay right now for nearly $1,000 a piece. Brigham sums it up: "When it boils down to it, if you make a nice product that people need or want, it tends to find an audience."

COOKING

Schools of culinary arts carry considerable weight, and sure, getting a job in many restaurants may require such pedigree. However, there is still a way in for the determined maker who loves the kitchen to turn this into a serious hobby and even live off their passion for pastries.

Learn the Limits

First, you have to jump in! "Go with a two pronged approach: research and practice," says celebrity chef and James Beard-nominated food writer Allison Robicelli. "You can't learn to bake unless you're actually baking, and the more you do it the better you get."

Follow all the channels of people that inspire you. Always hunt for something new to try, in order to expand your skill set. Remember though, everything you're watching is edited, even if they include some fumbles, they don't include the true feeling of hard work, and the crushing defeat of a collapsed macaron with no feet.

For small nibbles of skills, try tackling something that you have never tried, but has maybe one new technique in it. You've never

Fine Root - finerootsf.com, Knife & Saw - theknifeandsaw.com, Natalia Klenova - Adobe Stock

CALEB KRAFT is senior editor for *Make:*. He's been learning various crafting skills all his life and seriously loves the endorphin rush of going from completely incompetent to barely passable at a new thing.

Should You Apprentice?

Allison Robicelli: "This is what I did, and it saved me the $35,000 it would have cost to go to culinary school. If you walk into a place and ask if you can work for free, it's unlikely anyone will say no. Just make sure you set solid terms so you're not taken advantage of. I'd say six weeks tops. Then find another place you can apprentice at where you can expand your skill set. Ask tons of questions, and be ambitious! If you're taking your education into your own hands, there's no room for shyness."

Chris Brigham: "When I decided to make the career change and get into woodworking, I went back and forth between going to school and apprenticing. I decided apprenticing was best because you get thrown right into it and you learn from someone who is not only teaching you the craft but at least in my case, also how to run a business which was beneficial. I also didn't have to pay a bunch of money to go to school which was nice because you don't do woodworking for the big checks. I was willing and able to make almost no money to work and was lucky enough to find a really great mentor in Derek Chen of Council Design. This was another very lucky break; I was able to apprentice for someone whose work I really respected and was similar to what I wanted to do and he was just a great person and teacher as well."

whipped a mousse? Give it a try! Hunt out failure and get to know it. If every person you watch on YouTube says "don't over stir!" find out why. Stir till your arms hurt, and witness the results. You'll often learn a lot from the exercise, such as learning to recognize the point of no return.

Building a Base

You can easily dip your toes in the pool of baking for money. "Offer to do bake sales in your community, sign up for fairs and festivals, donate to local charity events," Robicelli says. Have a weekend get-together where you serve a limited run of the fanciest pastries you can concoct with your skills. You may be surprised that your friends are willing to pay a couple bucks in order to help cover the costs and time, and of course to delight in the results of your hard work.

The next step is to branch out. You may find that your family and friends gush over one thing in particular. You can easily set up a store on Etsy to sell it! Yes, you can sell food baked in your kitchen on Etsy. You'll need to do your homework to find what is required in terms of your local laws but a quick search will show you that many are, in fact, doing quite well on the platform, selling edible goods to fans worldwide.

Moving Up

At some point you may outgrow your home kitchen. Your KitchenAid mixer, once your workhorse, may be incapable of handling the constant abuse — or quantities — and it becomes time to step up to a 20-quart Hobart. At this point, you have many options. There are professional kitchen spaces you can rent in order to run your now thriving online business, or you could opt to open your own patisserie. Who knows, maybe you'll make a name for yourself like Jamie Oliver and end up on TV and owning a chain of restaurants. "Just be prepared to fail a lot at first," Robicelli cautions. "It's OK! Your mistakes will still be delicious."

KNITTING

Knitting has been around for too long to recount and has never really fallen from the public eye. Glance around on any

busy street and there's a good chance you'll see knitwear of some type. Sure, the majority of it may be manufactured in bulk by hulking machines, but the skilled knitter is alive and well in 2018, with a vibrant community for support. "There will always be demand for creative fiber artists regardless of formal or informal training for the industry and anything from art, fashion, architecture and emerging fields," says Victoria Pawlik, founder of the Electronic + Textile Institute Berlin.

Starting Off

Knitting and its related disciplines are quite diverse — pick an area that feels interesting to you and dive in to learn the lingo. "Some professionals guide new knitters towards a scarf because it covers how to start a piece and the two basic steps to knitting, as well as how to end the knitting," suggests professional designer Jolene Mosley. "Other professionals suggest a new knitter to select a project that is easy that interests them."

Either route will give a foundation in the language of knitting. "There are basics the new knitter needs to understand so that they can continue to learn and know what to ask when they run into a problem," Mosley says. "They should know the terms Cast on (CO), knit (K), purl (P) and Bind off (BO). The rest of the terms build off of those foundation terms and can be learned as they are needed."

As for where to get started with the fine arts of fiber, Mosley notes the options are as broad as they are traditional: "Skills are passed down generation to generation," she says. "They are taught in clubs or classes, they can be taught by friends or even strangers on the commuter train, they are taught by peers or by self, they are supported with workshops, guilds, classes and other ways as well."

Tying it Together

Once you've put time into learning the craft, there are plenty of opportunities to make a little cash with your hobby. "People who sell finished items can be generally found in shops and in fairs selling their items," Mosley says. "Designers do sell their patterns online and in printed format or even in books as the sole designer or as a group of designs. There are also knitters who sell tools that they make and can include needles and yarns."

Mosley adds, "People love to know the story behind a piece or of the person who created the piece. Share, share, share! Let people know what inspired the work, what thought went into making the process happen, and what you are about. The value of someone's time is what sells the item, otherwise it is all just twisted yarn."
— *Caleb Kraft*

TIP: Mixing pop culture into your designs can easily put your creations into common searches on Etsy, allowing for a healthy amount of goods sold.

The Cosplay Gateway

If there's one hobby that can help you master just about every maker skill, it's cosplay. Traditional **costume making**, **digital design and fabrication**, and **electronic prototyping** all come into play, as much or as little as a cosplayer sees fit. A global industry, cosplay continues to grow in popularity, offers a fantastic social network, and is a great area to gather a wealth of skills that can be applied to a wide variety of fields.

TEXTILES

Fabric is an unforgiving medium to work in. Not only are you sculpting a 2D object onto a 3D frame, but, from bias to weft, there are all sorts of rules to fabric crafting that aren't immediately apparent. It becomes even more challenging in cosplay when trying to create something like an animated character, where the costume never had to adhere to real physics.

"I do a fair bit of sewing," says Shawn Thorsson, master costume builder and author of *Make: Props and Costume Armor*. "I have a normal, modern, plastic sewing machine with a bunch of settings I don't even pretend to understand and a beastly, heavy, antique industrial sewing machine that will punch through half an inch of layered leather."

Make no mistake, sewing is hard. The good news is you can still learn enough on your own to build a project and determine if you have a love for the craft. "I make my living building costumes. The irony is that I absolutely hate sewing," says Thorsson. "I'm completely self taught, so I don't know any of the actual terminology. I just take garments apart and reverse engineer the assembly."

SEWING BASICS

The library is a wonderful resource for everything from historical clothing research, to sewing practices, to pattern drafting guides. Some libraries even lend patterns, or run exchange programs, and host clubs and meetups around different fabric arts. There also are lots of YouTube tutorials and online resources for workshops on different fabrics, styles, or eras. Your local fabric store is an excellent place to get real-world guidance on learning new techniques, offering classes and advice. Remember to check the

1

remnants bin for low-cost embellishment and trim options. Most fabric stores have annual sales with great deals, planning out your yearly con schedule around these is a great way to save some cash.

Fabric is only half the battle, often you will need to duplicate things like armor. Learning how to work with materials like stretch vinyl and leather will increase the kinds of items you can create. EVA foam is another important material to master, you will need to learn new ways of attaching and finishing. The internet is a valuable resource here, websites like Kamui Cosplay (kamuicosplay.com) and Punished Props (punishedprops.com) have detailed tutorials on working with these materials, as well as several books.

Many makerspaces do sewing meetups, and host workshops. This is a valuable way to make real-life connections. Look for your local conventions too. Often there will be workshops or panels on a cosplay track, which can teach you new methods, tips, tricks, and provide inspiration.

CAD AND 3D PRINTING

Digital fabrication has introduced wonderful efficiencies into the world of props, costumes, and fashion. "From the moment I could find affordable options, I've been using digital fabrication techniques such as 3D printing and CNC carving to make parts and prototypes for almost all of my builds," says Thorsson. "I still enjoy crafting and sculpting by hand, but it's hard to beat the speed and precision of having robots to do some of the work for me. At this point there are tons of usable options for free modeling software and countless online tutorials to get yourself started."

Fashion-tech designer Anouk Wipprecht, who is known for her elaborate 3D-printed dresses, echoes these sentiments: "I sculpted and carved by hand with hand tools, using A+B and airplane epoxies. Now with one click of a button you can repeat any complicated geometry and make it symmetrical, instead of sculpting for hours and hours."

But even without these design skills you can immediately employ your printer as a prop-making tool, as the internet abounds with files for accessories for all kinds of characters. After learning the print process, you can start experimenting with remixing

1. Shawn Thorsson's Sole Survivor power armor from *Fallout 4*.
2. Saura Naderi's articulated robot dress.

HEP SVADJA is the Product Manager for Adobe Capture, and was previously the photographer for *Make:*. In her spare time she is a space enthusiast, metal fabricator, and *Godzilla* fangirl.

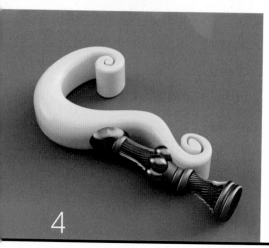

files to make your own customizations.

"One of the best parts about digital builds is the wealth of 3D models that are available online," says Thorsson. "If you're trying to dress up as something with even a modest fan following, chances are there's a forum or a website where someone has already posted a ready-to-print model of whatever pieces you need."

DIGITAL REPOSITORIES AND RESOURCES

The Replica Prop Forum: therpf.com
Thingiverse: thingiverse.com
My Mini Factory: myminifactory.com
Pinshape: pinshape.com
Cults: cults3d.com
Do3D: do3d.com
Yeggi: yeggi.com

There are plenty of choices for 3D design platforms to learn, and several of them have free or low-cost plans for makers. **Tinkercad** is a free web-based design program that is easy to learn yet robust enough to do serious designs. **Fusion 360** is a paid program, but they offer a free license for hobbyists and startups. Both of these have excellent easy-to-follow tutorials. SketchUp offers a free non-commercial version called **SketchUp Make** where you can learn the basics that are translatable to SketchUp Pro.

"In my shop I use every kind of free

software I can get a hold of," says Thorsson. "I get a lot of use out of SketchUp, Blender, MakeHuman (for when I need rough human bodies for digital test-fitting), and an outdated copy of Netfabb for when I want to verify that models are watertight and ready for printing or carving. They're all fun to tinker with and each one of them has significant online communities with tutorials and users who can answer questions while you're learning."

MICROELECTRONICS

So many cosplay ideas can be executed with just fabrication techniques, but what happens when you want to add effects, like having your Infinity Stones actually glow, or your lightsaber make noises when you swing it and hit things? The answer lies in the wonderful world of maker electronics.

"Microcontrollers are so cool," says Saura Naderi, whose robot dress collaboration stole the show at this year's Maker Faire Bay Area. "I fell in love with them in my senior design class in engineering school building a laser fire and control unit. I taught myself on an STK500 and it's gotten significantly easier to enter this realm since."

There are lots of platforms and communities geared to beginners, with tutorials and examples that can be leveraged for most project ideas. A bonus: You'll be learning how boards work the best way possible — by jumping in and doing it. "Anything that gets you practicing coding

5

6

in microcontrollers is helpful," Naderi says. "You'll naturally hit boundaries and appreciate why things like quad processing is rad."

GET TO WORK

Hardware development skills are in huge demand these days, but not just in tech. Schools, makerspaces, and learning programs all need teachers. Learning how to write about electronics by recording your build can be applicable to tech writing and documentation. Hardware fluency can even be helpful in the art field doing technical illustration.

If the hardware and software bug bites a cosplayer hard enough, they might even make it a full-time job. This is where having a portfolio of completed builds comes in handy, acting as a sort of resume of hardware projects you have seen through from start to finish. Applying for entry-level technical positions with this kind of diverse background will give you a leg up on the competition. "In reviewing a candidate, I tend to seek out things that show an intrinsic motivation to experiment, break things, and make technology part of your lifestyle," says Matthew Chambers, an infosec industry professional in charge of hiring. "It's about cultural fit, finding out about a candidate's hobbies, and gauging their curiosity about technology."

— Hep Svadja ⊘

Pick a Brain

In cosplay and beyond, the question of "which board" will drive which environment you use. Many start with the **Arduino Uno**, because of its legacy, its dedicated community, and the immense amount of existing projects for you to draw on or adapt. The **Circuit Playground** was designed as a learning tool so it has lots of integrated sensors, making it a smart choice for diving right in without needing to solder. There are a lot of small footprint boards on all kinds of programming platforms, perfect for discreetly hiding in your designs. Boards like the **Gemma** are designed to be sewn in, while other boards like the **Teensy** are great for nesting inside 3D prints. Our Guide to Boards (makezine. com/comparison/boards) is a useful resource for sorting your hardware decisions.

For those who do their best prototyping virtually, 123D Circuits is now included in Tinkercad. There is a virtual circuit system, so you can test your designs before building them in the real world. The only microcontrollers supported so far are the Arduino Uno and the ATtiny, but there are plenty of starter builds and components around those two for you to experiment with.

1. Awin Sidelet's hammer from the video game *Atelier Escha & Logy: Alchemists of the Dusk Sky*, fabricated by Fiora Aeterna.
2. Subject Delta Big Daddy and his Big Sister from *BioShock 2*, at SDCC.
3. Laphicet's compass from the video game *Tales of Berseria*, fabricated by Fiora Aeterna.
4. 3D render for the top of Shallotte's staff, from the game *Atelier Shallie: Alchemists of the Dusk Sea*, modeled by Fiora Aeterna.
5. Jacquie Skellington and her man Sal at SDCC with their dog Zero. Genderswap cosplay is a way to increase creativity when representing characters.
6. Hanzo and Genji cosplay from the video game *Overwatch*.

Commission It

Once you're up and running, you may find people asking you for commissions right off the bat. When determining your pricing make sure you do the cost analysis accounting for your working time, not just cost of materials. If you aren't ready to do full outfits yet, you can always concentrate on easy to churn out items like leather belts or fabric gauntlets.

Sewing cosplay costumes is one way to build a business, but often those same execution skills can be used to create sellable items for proms and weddings. Selling online means being able to serve to clients in areas that can do wedding season year round. Targeted Facebook and Instagram marketing will help you get your brand to the right eyes. Just make sure you plan well, so that your wedding deliveries don't overlap with con season.

Selling sewing supplies can be a smart business move. Marketing fat quarter lots of unique fabrics to quilters is one idea. If you have access to an embroidery machine there is good business in the custom applique market. Notions are another money maker, such as rainbow sets of zippers in different sizes, or color-matched selections of buttons and fasteners. Buy the supplies wholesale and then repackage as curated kits on Etsy.

Another great way to earn money is by selling patterns online. Etsy, Craftsy, Patternfish, and other similar platforms are good for this. Many people don't have the time to conceptualize and plan how they want to produce something, even if they have the skills for it. Pattern selling is a good way to leverage ideas you have already invested in, without a lot of work, leaving you free to focus on new ones.

Programmed for Success

BLACK GIRLS CODE is bringing skills — and a diverse voice — to technology's future

Written by Hep Svadja

HEP SVADJA is the Product Manager for Adobe Capture, and was previously the photographer for *Make:*. In her spare time she is a space enthusiast, metal fabricator, and *Godzilla* fangirl.

Kimberly Bryant

Kimberly Bryant never planned on starting a technology learning revolution. But when her 12-year-old daughter, Kai, expressed an interest in computer programming and found the male-dominated offerings disappointing, Bryant stepped up to fill the void.

Initially Bryant was looking to bring a small group of Kai's friends together for the weekend and teach them how to code. "And we found with that very first pilot class that there was a need that wasn't being met in the social sector around introducing these girls to computer science," she says.

Inspired by her experience, Bryant launched Black Girls Code in April of 2011, with a mission to teach young teen and pre-teen women of color programming skills via access to technology, workshops, classes, and community support. Seven years later, the nonprofit organization, headquartered in Oakland, California and New York, has helped more than 10,000 students and established programs in 13 states as well as South Africa. Girls not only get access to software engineering education, but also experiment with concepts like web and game design, app development, 3D printing, robotics, and VR. Black Girls Code continues to expand with new opportunities, such as their recently opened tech exploration lab in Google's New York offices.

An Unlikely Career Path

Bryant grew up in Memphis, Tennessee, in the 1970s — an era when women, especially women of color, were not considering computer science. "We were not expected to be scientists at all," she says.

Middle school courses sparked a love of sciences. "I was really just kind of enthralled with the things that made biology and biochemistry come to life," Bryant says.

In high school her guidance counselors noticed her math and science abilities. They encouraged her to explore engineering at Vanderbilt University, a field she didn't even know existed, involving gadgets and machinery that were generally off limits to a young woman. "It was very scary, learning how these things worked, and how I would actually be coding them and installing them," she says. "But it was also very exciting because when I started college it was really at the birth of the computer science and personal computer age, [we were] starting to ride that wave of learning

about this next technical revolution." Bryant went on to manage systems in heavy engineering in the manufacturing industry before doing project management for biotech giants Genentech and Novartis.

Building a Support System

Bryant believes banding together is imperative to success. While at Vanderbilt, a conversation with an electrical engineer upperclasswoman grew into a mentorship, as well as a network. "From that encounter I was able to connect with other communities of computer scientists and engineers, who were other students of color, and that is how I was able to survive and even thrive in my four years of college," she says.

Creating such a support system for girls like her daughter, so that they "could really find this tribe of other young women who have similar interests," was a big part of creating Black Girls Code, Bryant says. "I think especially when you're a woman in a male dominated industry, it's vitally important that we have a community of folks around us who have similar experiences, similar perspectives, who are all going on this journey together."

Community is what drives the organization — from the volunteers who come in and teach the girls on the weekends, to the parents who bring the kids to the workshops, to the educators who introduce the girls to Black Girls Code's programs from their school districts and from their classrooms. Getting involved is as simple as tapping into their resources online, going to blackgirlscode.com and filling out a volunteer application, or asking your company to get involved as a sponsor of one of their workshops. "Just reach out,"

Bryant says, "figure out a way that you can use your skills and your talents to help these girls learn and really grow into leaders."

A New Narrative

Silicon Valley still struggles with diversity, despite close proximity to Oakland and San Jose, both with large communities of color. As the tech industry moves into the East Bay, Oakland is doing better than average with building diverse workforces. According to the Oakland Chamber of Commerce "Tech Trends Report," in 2015 Oakland's tech sector was 11.7% African American, compared with 1.9% in San Francisco and 1.3% in San Jose. This is thanks in part to programs such as Black Girls Code, The Hidden Genius Project, and Hack the Hood, which all encourage diverse youth communities to explore and acquire tech skills.

Bryant emphasizes why Black Girls Code is so important for communities of color: "When we think of a computer scientist we don't often think of a woman of color," she says. "We certainly don't think of a black woman. We think of folks that don't look anything like me." Black Girls Code is trying to shift that narrative. "The work that Black Girls Code is doing is revolutionary in that we're saying yes, girls of color, black girls in particular, are coders. They're technologists and they are the future leaders of technology. That's a radical statement and something that folks have had to wrap their heads around, this new narrative for what a technologist and computer scientist looks like."

This is increasingly important going into a world where much of our lives are managed by technology. "We're doing things on an iPhone or a laptop that were never thought of even in my generation when I graduated," Bryant says. But that is just the beginning of what technology and the field of engineering is going to be capable of in the future. "I think it's vitally important that as we go into this next industrial-tech revolution, that we have diverse creators around the table," she says. "I think that it's important that we not only have women sitting at the table, but that we also have people of color who are creating, because the demographics of our world are very diverse. If we don't involve more voices in the very first step, of what technology is created and how it is used, I think we'll miss the boat in terms of how we'll use these innovative technologies that are on the horizon." ◉

Hep Svadja, Black Girls Code

Head of the Class

IRVINGTON HIGH SCHOOL'S MAKER SPACE
equips students for the future
Written by Clair Whitmer

CLAIR WHITMER is head of digital product development at *Make:* and led the launch of Maker Share.

See the Irvington High School's Maker Space work in this Maker Share showcase: makershare.com/showcases/berbawy-makers-class-act

Want to create a program like Irvington High's in your learning community? Sign up for our new *Make:* Workshop "How to Start a School Makerspace" with educator Adam Kemp.

School makerspaces are one of the hottest trends in educational technology. But beyond the label, there are a range of opinions about how to go about setting one up and assessing its impact, especially how maker education fits into standards-based expectations. Two common questions are whether or not schools should have a dedicated makerspace or distribute the equipment across the school. Administrators also debate if the goal is to reinforce the curriculum with project-based learning or to challenge what counts as learning in the first place.

Kristin Berbawy, teacher and founder of Irvington High's Maker Space, seems liberated from too much fussing over theory; she is driven more by her conviction that kids will create cool things if you give them the tools and get out of their way. The creation of her makerspace/classroom in Fremont, California — now equipped with a laser cutter, a CNC router, a vinyl cutter, 16 3D printers, and a mini-warehouse of electronics components — has required Berbawy to cross a 5-year obstacle course of networking, bartering, and evangelizing for her kids, many of whom have stuck with her for all four years of high school.

"Ms. B. talks to people and makes friends," sums up Aditya Rathod, 17. He came to Maker Faire this year — Irvington High's fifth time exhibiting there — with a version of Pac-Man using donated Oculus VR gear. One of his best moments at the Faire was meeting and getting tips from an Oculus engineer who noticed his project.

Berbawy teaches three subjects: Intro to Engineering Design, Principles of Engineering, and Intro to Computer Science. But "networking" may the most important skill that she imparts. As with that Oculus, the program has been funded by grants and donations and Berbawy has taken full advantage of her proximity to Silicon Valley.

She prepares all year to bring her students to Maker Faire and the kids are proud of being one of the biggest and boldest high school groups to exhibit. "We felt special at Maker Faire," says Shruthi Somasundaram, 17. "We had some of the coolest projects." Somasundaram brought her Erase-a-Bot, a whiteboard-erasing robot inspired by her math teacher, to Maker Faire with partner Shubhi Jain.

Portfolios
The kids also share a genuine commitment

to open source technology and a desire to share what they've made. "If you make something and can't communicate about it out to the world, then what's the point in making it?" says Rathod.

To reinforce this concept, for example, Berbawy required this year that all the kids create their own Maker Portfolios on Maker Share and put a QR code to that project on posters displayed in their booth at Maker Faire. Berbawy focuses on helping her students develop written and visual documentation skills, relying on the pipeline of student mentors she's established to man the makerspace and assist with the hands-on learning.

Senior Cody Pappa, 18, says that's how he gained experience with the equipment. "I fell in love with the machines and the older guys taught me how to really use them. I thought I knew how to solder ... but I didn't know how to solder," he says with a laugh. Pappa feels the most important thing he's learned in class is "iterative engineering." He says Ms. Berbawy makes them figure out why something failed and then try it again. Pappa is headed to San Diego State this fall to study mechanical engineering.

Teaching Teachers

Berbawy's next challenge is sharing what she knows with a broader "learning community" of educators. She's traveled around to feeder schools with a 3D printer on a cart to recruit incoming students. But she's still working to increase the diversity of her classroom: there are fewer girls than boys and much fewer African-American and Hispanic kids than whites or Asians. She believes the key is to reach them early. "By the time they get to high school, it's too late. They already think they can't and while thinking they can't, they get involved in other things," she says.

So Berbawy has now turned her energy from organizing the kids to organizing her fellow teachers: she's trying to build her own learning community in her district to share her knowledge of the equipment and teaching techniques. She dreams of turning her makerspace into a lending library for other teachers and schools.

In the meantime, her high school students are entering the world with an enviable confidence, fully capable of demonstrating their skills, undaunted by small setbacks, and comfortable advocating for themselves and their projects. Makers. ◎

Kristin Berbawy

Berbawy, Vasant Chalemcheria, and Sai Kesari

Aditya Rathod and Sidharth Khabiya's VRCade

Shruthi Somasundaram and Shubhi Jain's Erase-a-Bot

Cody Pappa loads 3D printer filament

Maker Spotlight: Brandon Quimson

Hep Svadja

One of the outstanding seniors in Berbawy's program is **Brandon Quimson**, 18. He brought both his Daft Punk-inspired and his Marshmello look-alike helmets to Maker Faire this year. "My Marshmello head might be even better than his," says Quimson, referring to the EDM producer and DJ.

Quimson is nothing if not confident. He just won the NCS Meet of Champions in pole vaulting, and he's off in the fall to California State University, Sacramento to join the track and field team and take his shot at competing at the Olympics. Oh, and he's going to study engineering too.

Quimson was diagnosed with dyslexia and ADD in fourth grade. By the time he met Berbawy, he'd already settled into a home-schooling program taught by his mother. Berbawy invited him to join her class anyway, negotiating the normal admin channels so that he could be part of her program. "Just come and we'll make it work," is what Quimson says Berbawy said.

Berbawy gave him both a social group and goals like Maker Faire that helped him turn what at first was diagnosed as a disability into a strength. "Due to ADHD, I can hyper-focus. When I get a new project, I forget to eat or sleep," he says. "I'm dedicated. I try not to half-ass my projects or my workouts. I trust the process and I know there's no instant gratification."

He even has a job doing "R&D and customer service" at drone manufacturer Mota. "Everyone has their cross to bear and mine is dyslexia. I just wish other kids had the opportunities I've had."

The Bright Stuff

Five tips to help **gifted kids succeed** in the classroom
Written by Michelle Muratori

MICHELLE MURATORI, PH.D., is a senior counselor/researcher at the Johns Hopkins Center for Talented Youth and a faculty associate at the Johns Hopkins School of Education.

Becca Henry

If you have a smart kid, someone who really loves to learn, school should be a breeze — a place filled with engaging learning opportunities and a supportive community of teachers and classmates.

But in my 15 years of working with hundreds of academically advanced pre-college students and their families at the Johns Hopkins Center for Talented Youth (cty.jhu.edu), I've learned that for many bright children, school poses a number of academic and social challenges, which can take their toll not only on academic performance, but on social and emotional well-being.

Fortunately, parents and educators can play an important role in helping academically talented kids make the most of their school experience. Here are a few examples of how to help bright children succeed in the classroom this year:

Your seventh-grader loves math and science but their report card says otherwise.
It could be that your child is under-challenged at school and may need a more appropriate level of learning. Explore with their school how they can be challenged not just through subject acceleration but with enrichment activities like independent projects that allow them to pursue a topic in depth.

Your kid who loves to learn says they are bored in school.
Engaging in hands-on projects can help young learners develop interests, deepen knowledge, and bring the material they are studying to life. Learning through making something can inspire students and help them feel engaged in the process and more connected to what they are learning.

Your bright fifth-grader says they hate school because they have no friends.
Help your child connect with peers over a shared interest. If a child loves math, encourage them to start or join a math club at school, or explore academic activities outside of school, such as math competitions and summer programs.

Your aspiring novelist has a meltdown when their latest book report gets a B.
It's not uncommon for bright students to internalize extremely high standards, which can turn into immobilizing perfectionism. Help them rein in unrealistic standards by establishing boundaries, modeling flexibility, and understanding that no one is perfect.

Your sixth-grader, a World War II expert, regularly forgets to hand in social studies homework.
Executive functioning is a skill that can be learned. Talk to your child and their teacher and brainstorm some possible solutions, like setting a homework reminder alarm on their phone. By making children accountable and involved in the solution, they are more likely to follow through and be successful.

If your concerns persist about any of these issues, seek professional advice from a school or independent counselor.

Make Today.
Build Tomorrow.

A WORLD OF IDEAS:
SEE ALL THERE IS TO KNOW

www.dk.com

Dog Bowl
Written and photographed by Larry Cotton
Coffee Roaster

Built-in bowl spinner, bean stirrer, and heat gun achieve a perfectly even roast, automatically

LARRY COTTON has finally given up on doing anything earthshaking. He loves electronics, music and instruments, computers, birds, his dog, and wife — not necessarily in that order.

Home coffee roasting has become so popular that Googling the subject yields millions of hits, and rightfully so: the results can be amazing. There's just no other way to a fresher cup of coffee. And unroasted (green) beans are readily available from hundreds of sources online, and occasionally locally.

Here's a cheap, quick, and reliable machine that roasts a decent amount of coffee beautifully with only 5 main parts: two $9 Black & Decker screwdrivers, a $15 heat gun, a cheap power supply and, oh yeah, a (new) dog bowl.

Various incarnations of the heat gun/dog bowl technique are well documented online. This one takes the concept to the next level. It's easy to build in a day or two, quick to roast, and, best of all, the roasting process is automated, reliably exposing the entire surface of every bean to the same heat.

Cost for the main parts is under $50. Most other materials can be found in a reasonably well-equipped shop. The screwdrivers and power supply are available from Amazon, the heat gun from Harbor Freight, and the wonderful dog bowl from

TIME REQUIRED:
A Weekend

DIFFICULTY:
Easy

COST:
$50–$70

MATERIALS

» **Dog bowl, stainless steel with anti-skid ring** Gofetch 33.81oz, Walmart #553515742
» **Heat gun, 1,500W** Harbor Freight #96289. Other heat guns may work, but the mounting will change.
» **Cordless screwdrivers (2)** Black & Decker AS6NG, from Amazon or Target
» **Flat board, 12"×18"** such as 1×12 pine or ¾" plywood
» **Project plywood: ½", 2'×2'; and ¼", scrap**
» **Wood block, scrap**
» **Switches, SPST (2)** Gardner Bender GSW-25 or similar single-pole, single-throw switches
» **Small piece of ¹⁄₁₆" Formica** or similar material
» **Skate bearing, 608 type, 22mm OD / 8mm ID** aka fidget spinner bearing; can substitute four ⁵⁄₁₆"×¾"×¹⁄₁₆" washers
» **Flat-head bolt, ⁵⁄₁₆"-18 × 1" long**
» **Power supply/adapter, 115VAC / 6VDC / 3A** Amazon #B00P5P6ZBS
» **Solid copper wire, 12 gauge, about 18"**
» **Wood dowels: ⅜"×4", and ¼"×10"**
» **Hinge, 1½"×1", 4-screw mounting**
» **Aluminum strips, ¹⁄₁₆" thick: ½"×4", and 1½"×4"**
» **Aluminum sheet (flashing), 2⅝"×4"**
» **Spring**
» **Miscellaneous screws, washers, and nuts**
» **Wire-holding staples (3 or 4)**
» **Hookup wire, 22 gauge, about 4'** aka "telephone wire"
» **Heat-shrink tubing or electrical tape**
» **Mini desk fan (optional)** I used Mainstays #34136721, from Walmart.
» **Green coffee beans**

TOOLS

» **Hammers, regular and rubber**
» **Pliers and wire cutters**
» **Sheet metal cutter**
» **Scissors**
» **Hand screwdrivers**
» **Drill and bits, including ⅞" spade**
» **Cutting oil**
» **Countersink, ½" or ¾"**
» **Drill press (optional)**
» **Sander and sandpaper, 120- and 320-grit**
» **File and/or grinder**
» **Jigsaw** with wood and aluminum cutting blades
» **Band saw (optional)** with wood cutting blade
» **Soldering gun and electrical solder**
» **Dremel (optional)** with ¼" straight cutter
» **X-Acto knife** with new No. 11 blade
» **Vise**
» **Center punch**
» **Ruler, straightedges, circle templates**
» **Lots of sharp pencils with erasers**
» **Tape**

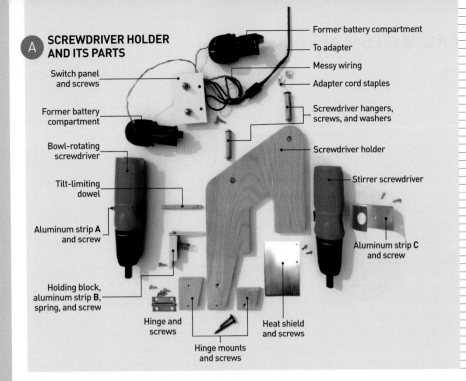

A SCREWDRIVER HOLDER AND ITS PARTS

Switch panel and screws
Former battery compartment
Former battery compartment
To adapter
Messy wiring
Adapter cord staples
Bowl-rotating screwdriver
Screwdriver hangers, screws, and washers
Screwdriver holder
Tilt-limiting dowel
Stirrer screwdriver
Aluminum strip **A** and screw
Aluminum strip **C** and screw
Holding block, aluminum strip **B**, spring, and screw
Hinge and screws
Heat shield and screws
Hinge mounts and screws

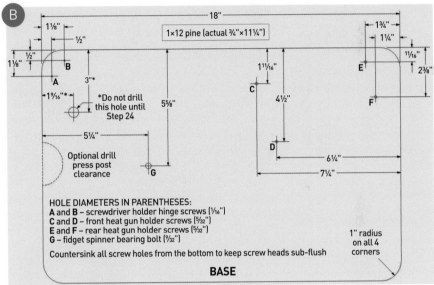

B

18"

1×12 pine (actual ¾"×11¼")

1⅛"
½"
1¾"
1¼"
½"
1⅛"
B
1¹¹⁄₁₆"
E
¹¹⁄₁₆"
2⅜"
A
3"*
*Do not drill this hole until Step 24
5⅝"
C
4½"
F
1⁹⁄₁₆"*
D
5¼"
Optional drill press post clearance
G
6¼"
7¼"

HOLE DIAMETERS IN PARENTHESES:
A and B – screwdriver holder hinge screws (¹⁄₁₆")
C and D – front heat gun holder screws (⁵⁄₃₂")
E and F – rear heat gun holder screws (⁵⁄₃₂")
G – fidget spinner bearing bolt (⁹⁄₃₂")

Countersink all screw holes from the bottom to keep screw heads sub-flush

1" radius on all 4 corners

BASE

Walmart. (It's wonderful because it has a rubber anti-skid ring and rounded corners.)

If you live close to a coffee roasting business, ask if they'll sell you some green beans. If not, Amazon will be happy to have some on your doorstep by the time you finish reading this article. OK, maybe not quite, but go ahead and order 5 pounds of unroasted Kenya AA and start building!

Figure **A** identifies most of the parts that will keep the beans constantly moving. This assembly will be mounted to a wood base, which is made first. If you want to paint or clear-coat the wood parts, do so before any construction.

MAKE THE WOODEN PARTS

1. Make the base from a flat 18" piece of 1×12 pine (¾"×11¼") or ¾" plywood.

Drill holes A–G as accurately as you can, following the *x-y* dimensions provided in Figure **B**.

Do not drill the hole that's marked with asterisks (*) until Step 24.

2. In the bottom of the base, countersink holes C, D, E, and F to fit flat-head #8×1½" wood screws. Countersink hole G so the ⁵⁄₁₆" bolt's flat head will be barely sub-flush with the bottom of the base. Thread the bolt into that hole. Put the skate bearing or fidget spinner bearing over the end of the bolt. If you can't find the specified bearing, use a stack of four ⁵⁄₁₆"×¾"×¹⁄₁₆" flat washers. The bolt must not protrude above the bearing (or washers). If it does, unscrew the bolt a bit, keeping its head sub-flush, and/or grind its end.

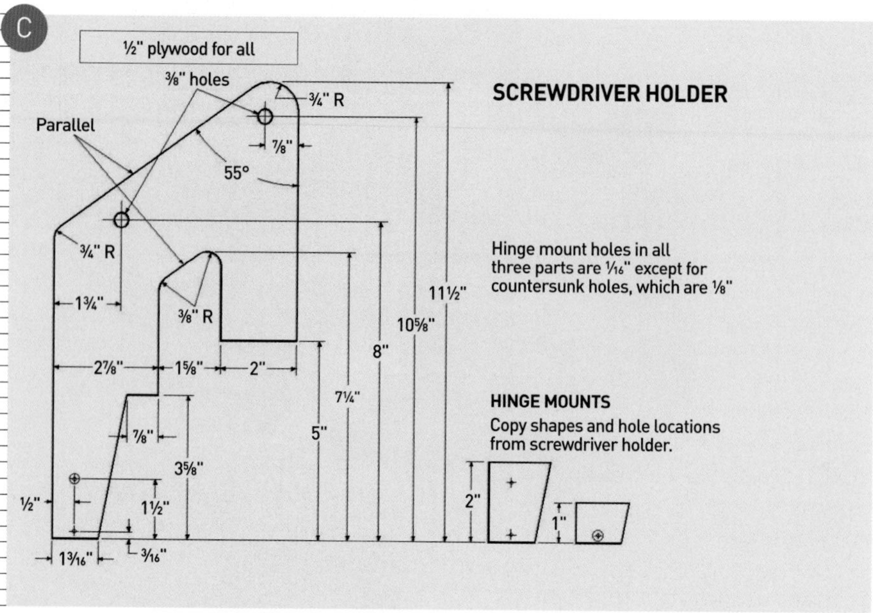
C
½" plywood for all
⅜" holes
Parallel
¾" R
⅞"
55°
SCREWDRIVER HOLDER
¾" R
1¾"
⅜" R
2⅞"
1⅝"
2"
11½"
10⅝"
8"
7¼"
5"
Hinge mount holes in all three parts are 1⁄16" except for countersunk holes, which are ⅛"
HINGE MOUNTS
Copy shapes and hole locations from screwdriver holder.
⅞"
3⅝"
½"
1½"
2"
1"
1 3⁄16"
3⁄16"

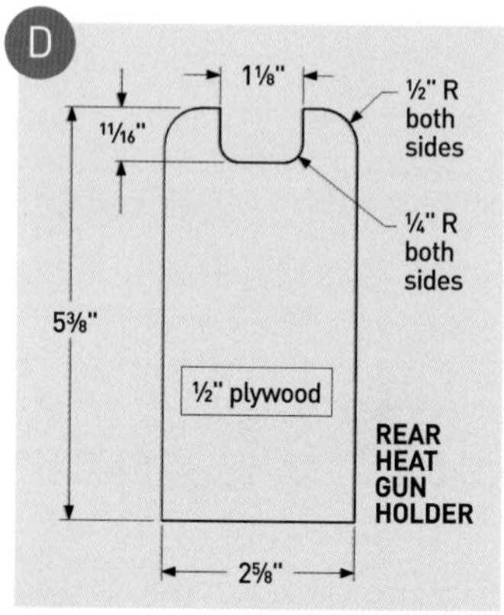
D
1⅛"
½" R both sides
11⁄16"
¼" R both sides
5⅜"
½" plywood
REAR HEAT GUN HOLDER
2⅝"

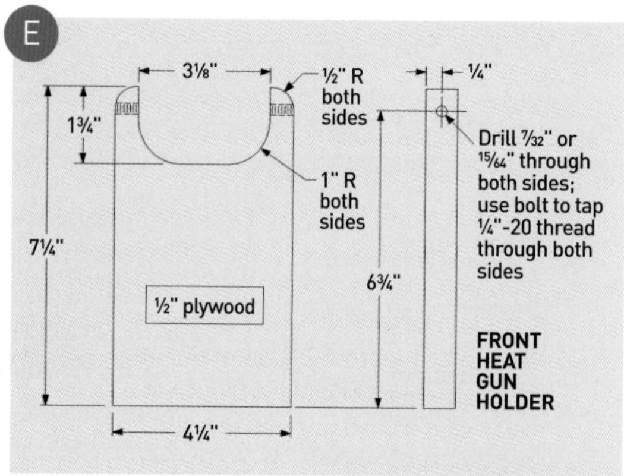
E
3⅛"
½" R both sides
¼"
1¾"
1" R both sides
Drill 7⁄32" or 15⁄64" through both sides; use bolt to tap ¼"-20 thread through both sides
7¼"
6¾"
½" plywood
FRONT HEAT GUN HOLDER
4¼"

If your base is not perfectly flat, attach two or three 2×2 stringers, 11¼" long, perpendicular to the base's long axis. Just don't cover any holes.

3. Make three more wood parts in this step and Steps 4 and 5 from ½" project plywood. Use a jigsaw or band saw to cut out the screwdriver holder and hinge mounts (Figure C). Holes are dimensioned to fit #8×1" and #8×1½" wood screws. Don't mount any wood parts to the base yet.

4. Make the rear heat gun holder (Figure D). Keep the cutout centered, and no deeper than 11⁄16".

5. Make the front heat gun holder (Figure E). Its cutout must also clear the housing. Ensure the two threaded holes line up with the mounting bosses on the heat gun. Sand all wood parts with 120- and 320-grit paper. Go ahead and install the two ¼"-20 × ¾" machine screws; these will lock the heat gun in place later.

BUILD THE ROASTING BOWL

6. See Figures F, G, and H for this step and Steps 7–10. Make the bearing surround from ¼" project plywood. On one side, slightly countersink the two 1⁄16" holes to clear the flat-head screw heads when they poke out of the bottom of the bowl.

The ⅞" hole fits the specified bearing. If you use washers instead, drill a ¾" hole. In either case, a spade bit is best for this task.

7. On the bottom surface of the bowl, place the surround with its small countersinks toward the bowl, and its big hole concentric with the machined circles in the bowl. This must be accurate so that the bowl won't wobble excessively side-to-side. (Up to 1⁄16" wobble is OK.) Tape it in place.

8. Using a new 1⁄16" bit and cutting oil, slowly drill through the surround's two 1⁄16" holes into and through the bowl bottom.

9. From the top of the bowl, gradually countersink the two holes just enough so that the heads of two #4×¼" flat-head screws will not pass through, but will be virtually flush. Attach the surround. If you countersink a hole too much, just drill and countersink another pair of holes; a couple extra holes won't affect the roasting

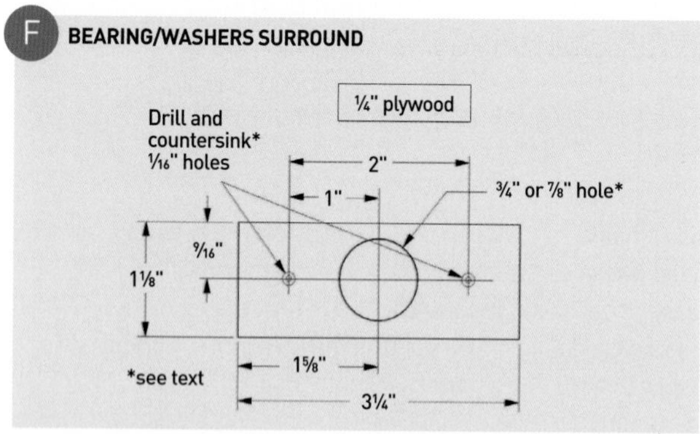
F
BEARING/WASHERS SURROUND
¼" plywood
Drill and countersink* 1⁄16" holes
2"
1"
¾" or ⅞" hole*
9⁄16"
1⅛"
*see text
1⅝"
3¼"

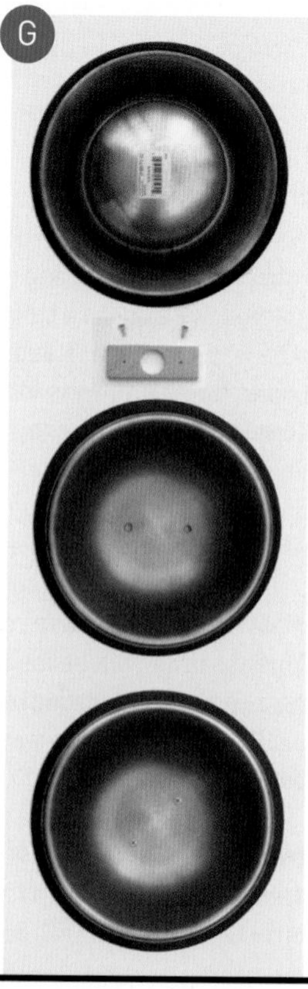
G

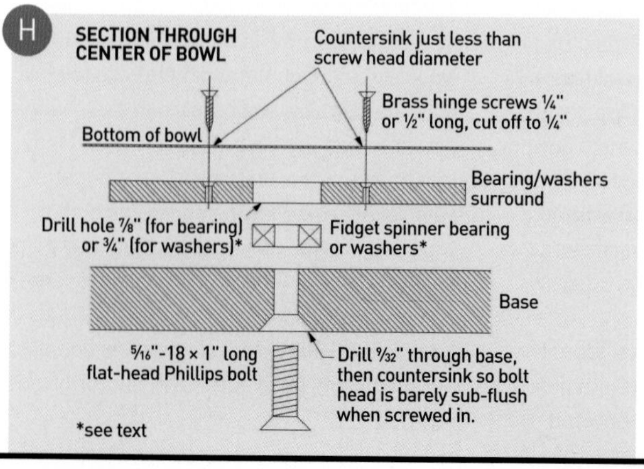
H
SECTION THROUGH CENTER OF BOWL
Countersink just less than screw head diameter
Brass hinge screws ¼" or ½" long, cut off to ¼"
Bottom of bowl
Bearing/washers surround
Drill hole ⅞" (for bearing) or ¾" (for washers)*
Fidget spinner bearing or washers*
Base
5⁄16"-18 × 1" long flat-head Phillips bolt
Drill 9⁄32" through base, then countersink so bolt head is barely sub-flush when screwed in.
*see text

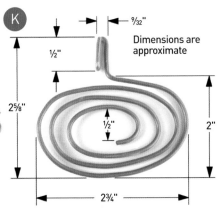

Dimensions are approximate

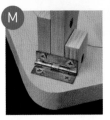

process. Those screw heads need to be flush so the bean stirrer doesn't hit them.

10. File, sand, or grind flush the points of the mounting screws if they protrude from the bottom of the surround (Figure **I**).

11. Place the bowl/surround assembly over the skate/fidget spinner bearing (or washers). It won't be easy to locate the first time, so draw a circle on your base around the bowl to speed the process later.

12. Give the bowl a 360° spin. If it wobbles side-to-side more than $1/16$", re-mount the surround more accurately to the bowl. The bowl's anti-skid ring will be driven by a screwdriver chuck (that normally holds a driver bit). A bit of vertical wobble is OK.

MAKE THE BEAN STIRRER

13. Make the stirrer from 12-gauge solid copper wire. I bought a few feet of 3-conductor cable and stripped about 18" of insulation from one of the conductors with an X-Acto knife and No. 11 blade.

Bend one wire end twice to yield three $1/2$" segments that will fit into the stirrer screwdriver's chuck (Figure **J**). Roll up a rough elliptical coil, which can bulge slightly at first, then flatten it (Figure **K**) and try to keep it symmetrical. This will probably need to be adjusted after you assemble the machine, so just set it aside.

MODIFY THE SCREWDRIVERS

14. Since the roaster is powered by a 6VDC power supply instead of batteries, the screwdriver battery compartments must be modified. But first test each screwdriver: install four AA batteries, ensure the spindle-lock collar is unlocked, then press each button for forward and reverse directions. Remove the batteries.

15. In each compartment, at the end and to one side of its centerline, drill a $1/8$" hole to allow two 22-gauge wires to pass. Run the wires through the hole and solder them to the terminals that connect to the screwdriver (Figure **L**). Plug the battery compartments back in.

MOUNT THE SCREWDRIVER HOLDER

16. Add the two hinge mounts (cut out in Step 3) to the sides of the screwdriver holder using #8×1" and #8×$1 1/2$" wood screws (Figure **M**).

17. To easily place and remove the bowl, the screwdriver holder swings out of the way on a $1 1/2$"-wide × 1" hinge. Mount the hinge with #6×$1/2$" wood screws so it lines up with the bottom of the screwdriver holder assembly. Attach that assembly with #6×$3/4$" flat-head wood screws using holes A and B you drilled into the base.

MOUNT THE SCREWDRIVERS

18. Hang the screwdrivers on their holder with two $3/8$" dia. × $1 5/8$" dowels per Figure **N**. Drill the dowel ends to take 4 small screws; add 4 washers to trap the screwdrivers. Run wires over the top of the screwdriver holder. The outermost screwdriver turns the bowl and the inner one stirs the beans.

19. Make three $1/16$"-thick aluminum strips (A, B, and C) per Figure **O**. Strips A and B work with the bowl-turning screwdriver; C does double duty retaining the stirring screwdriver and holding one of its buttons in. All three pieces can be cut from standard $1/16$"-thick aluminum extrusions. Bend strip C around a scrap piece of PVC pipe, or even one of the screwdrivers. A vise and rubber hammer can help.

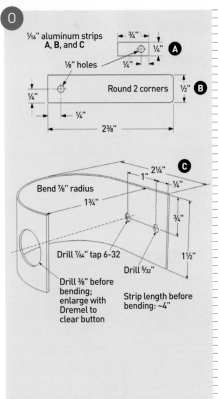

Larry Cotton

1/16" aluminum strips A, B, and C

1/8" holes

Round 2 corners

Bend 7/8" radius

Drill 7/64" tap 6-32

Drill 5/32"

Drill 3/8" before bending; enlarge with Dremel to clear button

Strip length before bending: ~4"

P

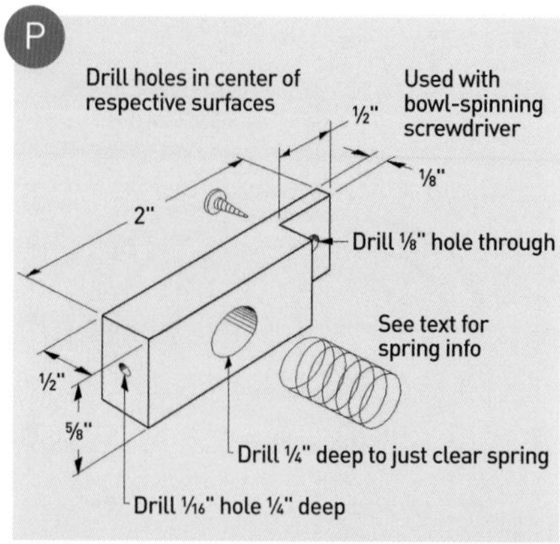
Drill holes in center of respective surfaces
Used with bowl-spinning screwdriver
1/2"
1/8"
2"
Drill 1/8" hole through
See text for spring info
1/2"
5/8"
Drill 1/4" deep to just clear spring
Drill 1/16" hole 1/4" deep

Q

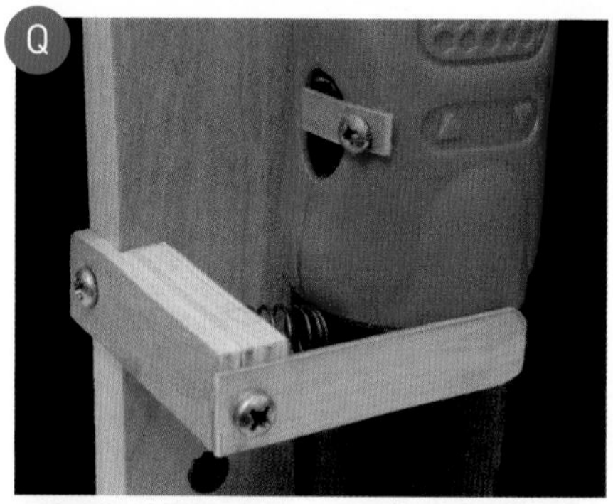

R

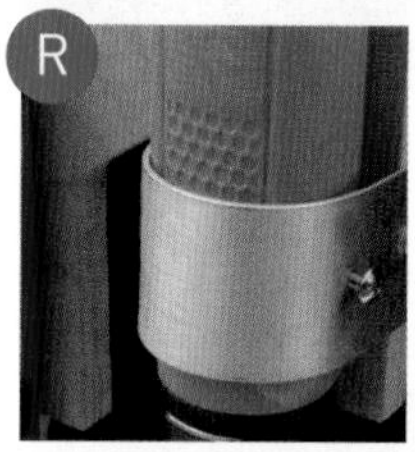

S

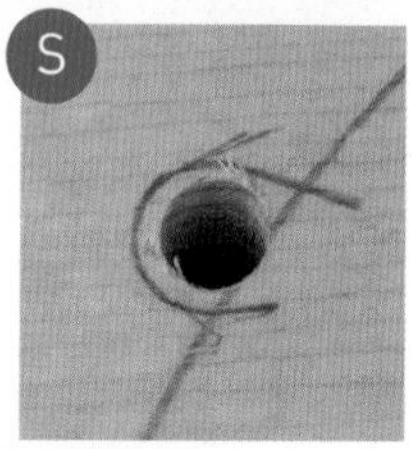

T

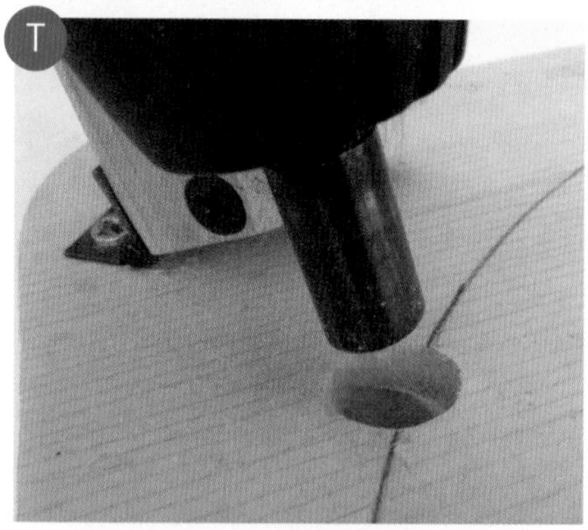

U

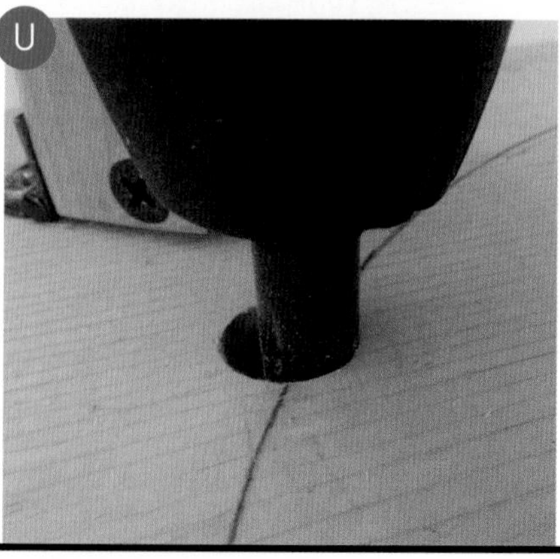

20. Make a small wood block (Figure P) to hold a compression spring and aluminum strip B. The spring must keep the bowl-turning screwdriver's chuck pressed against the bowl's anti-skid ring. When the bowl wobbles a bit, the chuck will follow it. Ideal spring dimensions are: 1/32" wire thickness, 7/8" long, 3/8" diameter, 10 coils.

Other springs will work as long as they provide a force of about 1/2lb to 1lb when compressed about 1/4". You could cut down a spring that's too long or stretch a spring that's too short.

21. Refer to Figure Q for this and the next step. Put a slight bend in aluminum strip A and drill a 1/16" hole into the bowl-turning screwdriver housing — just into the housing, not anything behind it. (Wrap tape on the drill bit as a stop.) Attach the strip to the screwdriver with a #4×1/4"-long sheet metal screw so the button is held in.

22. Using two #6×1/2" sheet metal screws, assemble the block, spring, and aluminum strip B, then attach the assembly to the screwdriver support. Ensure the spring applies pressure to the screwdriver so it will maintain contact with the bowl's rubber ring. The screwdriver must slightly — and freely — pivot about its hanging dowel.

23. Attach aluminum strip C to the screwdriver holder, and around the stirrer screwdriver, with a #6×1/2" sheet metal screw per Figure R. Add a #6-32 × 3/8" machine screw to hold one button in while the big hole in the strip clears the opposite button.

24. Hold the bowl-turning screwdriver roughly vertical, with its chuck sitting on the circle you drew around the bowl in Step 11. If the chuck is out of position, check your work — especially the hinge location — and for possible binding. Draw with pencil around the chuck's tip and drill a 3/8" hole (Figure S). This hole's locating dimensions in Figure C are approximate. The hole, for now, will be too small for the chuck to enter.

25. The chuck needs to drop about 1/8" below the top of the base so it will always engage the bowl's anti-skid ring. So enlarge the top of the hole with a round file or 1/4"-diameter Dremel cutter to clear the end of the chuck (Figures T and U). Install the bowl and rotate it 360°. Even with a bit of wobble, the chuck must always press on the bowl's anti-skid ring and not hit the base. Otherwise you'll have a bowl o' burnt beans.

FINISHING TOUCHES

26. Attach a heat shield for the bowl-turning screwdriver (Figure V). Cut it from thin aluminum sheet (flashing), 2 5/8"×4", and mount it to the screwdriver holder with two #6×1/2" sheet metal screws.

Also add about 4" of 1/4" dowel to the back edge of the screwdriver holder to keep the screwdrivers from tilting back too far.

27. Both screwdrivers should be vertical while running. If they lean in a bit too much and/or the chuck drops more than 1/4" below the base surface, add a small flat-head stop screw per Figure W.

WIRE IT UP

28. With the power supply disconnected, connect everything else following Figure X. My switch panel is plastic laminate ("Formica") about 2"×3", mounted with two screws; your mileage may vary depending on switches. Allow enough power supply wire to reach even when the screwdrivers are tilted back.

Before you solder anything, plug in the power supply and check each screwdriver's rotation direction (sequentially; your power supply will thank you). If a screwdriver doesn't run, check for power and that a reverse/forward button is firmly held in. Looking from the top, the bowl should turn clockwise and the stirrer counter-clockwise. Reverse wires if necessary.

Solder, and use heat-shrink tubing to cover bare connections. Staple the power supply wire to the base.

MOUNT THE HEAT GUN

29. You'll mount both heat gun holders with #8×1 1/2" flat-head wood screws. (Drywall screws work well.)

Mount the front holder to the base after drilling 1/16" pilot holes in its bottom using the corresponding holes and spacing on the base.

30. Similarly, mount the rear heat gun holder, but with only one screw (at first) so it pivots, allowing some alignment. Set the heat gun in both holders, then add the second screw.

31. Lightly tighten the two ¼"-20 screws into the heat gun bosses.

> **IMPORTANT:** The heat gun must be rigidly mounted, pointing below the stirrer screwdriver chuck and at the far rounded inside corner of the bowl where the beans will be agitated. If it's not pointing there, check your dimensions on the base and both heat gun holders.

NOW ROAST THOSE BEANS

Roasting a batch should take around 10–15 minutes. I've roasted coffee outside at 50°F (with no wind) and warmer with no problems.

1. Use the roaster in a dry, well-ventilated place without fire hazards, preferably outdoors.

2. Insert the stirrer into its chuck. You may have to tweak the chuck end of the copper wire so that it's difficult to push in; this should keep it from falling out. It must spin freely, just clearing the bowl's inner surfaces, but with no scraping.

If the stirrer occasionally falls out when swinging the screwdrivers back, drill a ¹⁄₁₆" hole in one side of the chuck: file a flat on one side of the chuck, and use cutting oil and a new bit. Then push a thick (¹⁄₁₆") bare retaining wire into the hole through the end of the stirrer's inverted U shape. Wrap the wire around the chuck a couple turns (Figure Y).

3. Before adding any green beans to the bowl, test your screwdrivers and heat gun one more time. Read the Harbor Freight heat gun manual. Note that the switch does not have a "cool" setting. Leave the heat gun off.

4. Easy to forget: When positioning the bowl-rotating screwdriver, push it out against its spring, then drop its chuck in its hole in the base and release, so the spring can push the screwdriver chuck against the dog bowl's anti-skid ring. Keep the anti-skid ring clean and free of any contamination, such as oil and dust.

5. Put 1 cup of green coffee beans into the bowl (Figure Z). With the heat gun off, switch both screwdrivers on (sequentially, remember?) for a minute. The bowl must turn slowly and reliably and the stirrer must robustly agitate the beans with no interference. Take your time with this step.

> **WARNING:** Do not switch on the heat gun until the bowl and stirrer have been turning with no interruption or interference for about a minute. Stationary coffee beans can catch fire very quickly! Always stay with your roaster until everything is turned off, and keep kids at a safe distance.

6. With both screwdrivers running, switch the heat gun to high. Beware the hot nozzle!

7. After a few minutes, the heat gun should blow a considerable amount of tan, light, and thin chaff from the beans as they expand. They'll begin to yellow, then gradually darken. You'll hear them crackle (aficionados call that the "first crack") as they expand and release oils. Also note the stirrer changing to lovely shades of red and purple as it's exposed to heat. No worries there.

8. When the beans are roasted to your liking and/or just starting to crack again ("second crack"), turn the heat gun off, but leave the screwdrivers running in position. Let the bowl and stirrer rotate for a few minutes in the cooler air until the bowl and beans cool enough to touch.

9. Swing the screwdrivers out of the way, remove the bowl, dump the beans into a container and seal it. Purists wait anywhere from 4 to 24 hours before grinding the coffee. Good luck with that! ◉

> **[+]** For the possibly impatient, I added a cooling fan to this rig to speed the cooling time between batches. To get the details, plus more roasting tips and resources, visit the project page online at makezine.com/go/dog-bowl-coffee-roaster.

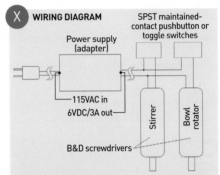
X WIRING DIAGRAM

Power supply (adapter)

115VAC in
6VDC/3A out

SPST maintained-contact pushbutton or toggle switches

Stirrer

Bowl rotator

B&D screwdrivers

Y

Z

Larry Cotton

TIME REQUIRED:
3–4 Hours

DIFFICULTY:
Intermediate

COST:
$60–$90

Eye of Newt

Written by Erin St. Blaine

For a charm of powerful trouble / Like a hell-broth boil and bubble!

Keep watch with a creepy, compact, animated eyeball. Put it in a wide-mouth jar and add it to your potion shelf, or attach a leather thong to wear it like a pendant around your neck. This guide is based on the Uncanny Eyes project by Phil Burgess, with a Halloween-y twist.

Before you start soldering, get all your software running and uploaded to your Teensy microcontroller. Getting the code loaded up first will make it easier to troubleshoot any soldering or build issues later on.

Software setup is fully covered in the Uncanny Eyes project at learn.adafruit.com/animated-electronic-eyes-using-teensy-3-1/software. And you'll find the hardware build at learn.adafruit.com/eye-of-newt.

CODE DOWNLOAD

Make sure you have installed everything listed below before moving on:
» Arduino IDE
» Teensyduino Installer
» Libraries (installed via the Arduino IDE and NOT the Teensyduino installer):
  » Adafruit_GFX
  » Adafruit_SSD1351
  » Adafruit_ST7735
» Python PIL Library (only if you want to add your own custom images)

Now download the project code from github.com/adafruit/Teensy3.1_Eyes/archive/master.zip. Inside you'll find a folder called *convert* that contains several different image folders and a Python script, and another folder called *uncannyEyes* that contains the Arduino sketch.

Open the sketch, *uncannyEyes.ino,* in the Arduino IDE and then make sure to select 72MHz as your CPU speed. (If your eye looks grainy, this could be your problem. It doesn't work right at the default CPU "overclock" speed.)

Upload the sketch to the Teensy as-is for testing, and make sure that it works before making changes.

Now look at the *uncannyEyes.ino* sketch. At the top you'll find several eye options.

Uncomment the `#include newtEye.h` line to turn on the newt eye option, and comment out the `#include defaultEye.h` line (Figure A). There can be only one!

This code defaults to rendering two eyes. Since we only have one eye, we can turn off the second one to make the code run faster. Just a few lines down, look for the `eyepins[]` array and comment out the second line within (Figure B) to turn off the right eye.

CUSTOMIZING THE GRAPHICS

I wanted an eyeball that looked as much like a real newt's eye as possible. I did an image search and found one I liked (Figure C).

Then I used Photoshop to "unroll" the eyeball so the software can draw it correctly. After some cropping, zooming, and judicious use of the Liquify filter, Figure D is what I ended up with.

The sclera (the white part of the eye) on a human (Figure E) looks really different from a reptilian eye. I wanted a more newt-like look, so I inverted the colors in Photoshop, then added a black circle to the center to keep the pupil dark. It took me several tries to get it right, but I'm really happy with the end result (Figure F). These images are included with the code download, and the process is explained thoroughly over at the Uncanny Eyes guide. Go nuts and create your own unique look.

EYE ORIENTATION

There's one more change we can make in the code to alter the orientation of the image. If your build comes out sideways or upside-down, and you want to rotate the eye to compensate, look for this line in the code (Figure G), at the very end of the `setup` function.

To rotate the eye 90°, change **(0x76)** to **(0x77)** or **(0x75)**. Or to rotate it 180°, use **(0x66)**. I personally like this eye rotated 180° degrees to upside-down from the original image. I think It makes the eye look like it's up to something crafty, which is really what I'm looking for in my Eye of Newt.

MATERIALS
» **Teensy 3.1 or 3.2 microcontroller**
» **OLED 16-bit color display** Adafruit #1431
» **Photoresistor, CdS** Adafruit #161
» **LiPoly battery charger** Adafruit #2124
» **LiPoly battery, 500mAh** Adafruit #1578
» **Slide switch, SPDT** Adafruit #805
» **Resistor, 10kΩ**
» **Half-sphere cabochon, 1.5" acrylic**
» **Hookup wire, solid core** multiple colors
» **Silicone stranded wire** multiple colors
» **Reptile print or Halloween fabric**
» **Necklace cord or wide-mouth jar**

TOOLS
» **Soldering iron and accessories**
» **Hot glue gun**
» **Scissors**
» **Needle and thread**
» **Gaffer's tape or duct tape**

ERIN ST. BLAINE is a fashion and LED artist based in the San Francisco Bay Area.

Ⓐ

```
#include <SPI.h>
#include <Adafruit_GFX.h>          // Core graphics lib for Adafruit displays
#include "logo.h"                   // For screen testing, OK to comment out
// Enable ONE of these #includes -- HUGE graphics tables for various eyes:
#include "defaultEye.h"             // Standard human-ish hazel eye
//#include "noScleraEye.h"          // Large iris, no sclera
//#include "dragonEye.h"            // Slit pupil fiery dragon/demon eye
//#include "goatEye.h"              // Horizontal pupil goat/Krampus eye
//#include "newtEye.h"              // Eye of newt
```

Ⓑ
```
eyePins_t eyePins[] = {
  {  9, 0 }, // LEFT EYE display-select and wink pins
//{ 10, 2 }, // RIGHT EYE display-select and wink pins
};
```

Ⓒ
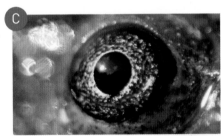

Ⓓ

Ⓔ    Ⓕ

Ⓖ
```
#else // OLED
  eye[0].display.writeCommand(SSD1351_CMD_SETREMAP);
  eye[0].display.writeData(0x76);
#endif
```

Supertrooper - Focusingonwildlife.com, Erin St. Blaine, Phil Burgess

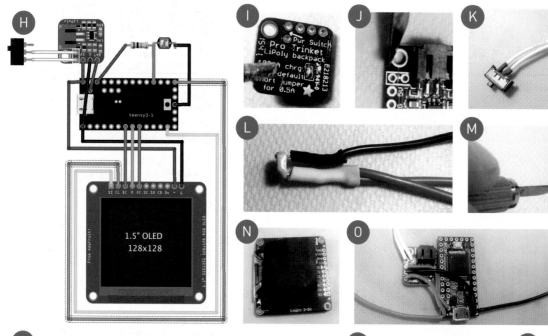

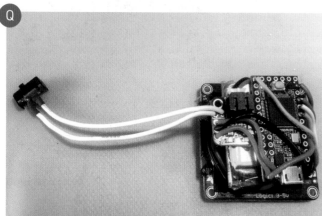

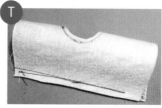

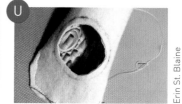

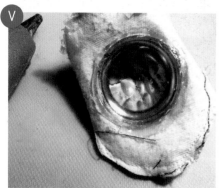

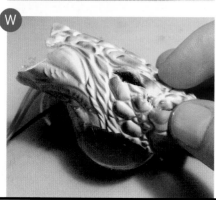

Erin St. Blaine

TROUBLESHOOTING

If you're having trouble, head over to the Uncanny Eyes guide and take a look at some of the troubleshooting ideas. If you see an eye on your display but it looks snowy and pixelated, check to be sure you've selected 72MHz as your CPU speed as noted before.

WIRING DIAGRAM

There are a lot of connections that need to be made (Figure H). Using a combination of solid core wire and stranded wire is the easiest way to get everything packed into as small a footprint as possible.

Color-coding is your friend here! Keep your power wires all red, and ground wires all black, and use a variety of colors for the other connections so you don't get confused. Write down the colors you used and the corresponding pins they connect with so you have a reference for soldering.

Teensy 3.1	OLED	Backpack Charger
Vin	+	BAT
G	G	G
USB		5V
7	DC	
8	Reset	
9	OC	
11	SI	
13	CL	
16		Resistor + Photocell
3.3V		Resistor
G		Photocell

ASSEMBLY

1. PREP YOUR CHARGER

Bridge the charge pad on the back with a blob of solder, to make your battery charge faster (Figure I).

Also cut the trace between the switch pads on the front to enable your on/off switch (Figure J).

2. PREP YOUR SWITCH

Trim the switch legs to about half their length. Solder 4" wires to the middle leg and one of the side legs, and cut off the other side leg. Secure the connections with heat-shrink tubing. Solder the two switch wires into the switch pads on the charger (Figure K).

3. PREP YOUR PHOTOCELL SENSOR

Trim your photocell's legs to about ⅓ of their length. Solder a stranded black wire to one leg, and two stranded colored wires to the other leg (the legs are interchangeable so it doesn't matter which is which). Cover each connection with heat-shrink (Figure L), then cover the whole photocell with larger heat-shrink, leaving only the top visible and uncovered.

4. PREP YOUR TEENSY AND DISPLAY

Cut the trace between the USB charging pads on the back of the Teensy (Figure M).

Place a large piece of thick tape (gaffer's tape or duct tape works great) over the back of the OLED display, carefully covering all the exposed components but leaving the solder hole labels visible (Figure N).

5. SOLDER POWER WIRES AND CHARGER

Using silicone stranded wire, solder two red wires into VIN and two black wires into G on your Teensy. We're using silicone stranded wire here because the solid core wires won't fit two-to-a-hole.

Set the charger next to the Teensy and solder a red solid core wire from Teensy's USB pin to the charger's 5V pin. Solder one of the stranded red wires to BAT and one of the stranded black wires to G (Figure O).

6. CONNECT PHOTOCELL AND DISPLAY

Solder various colors of solid core wires to the Teensy's pins 7, 8, 9, 11, and 13. You'll trim these to length later; for now just be sure they're at least a couple inches long.

Trim one leg of your resistor down and solder it into the 3.3V pin on the Teensy. Solder the other leg to one of the colored wires coming from your photocell sensor. Cover the whole resistor with heat-shrink.

Solder the other colored wire from the photocell into Teensy's pin 16, and solder the photocell's black wire into the GND pin next to the Teensy's reset button.

Place the Teensy and charger in line with

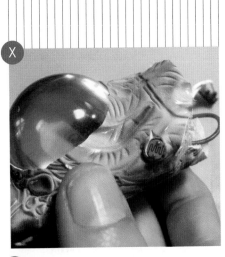

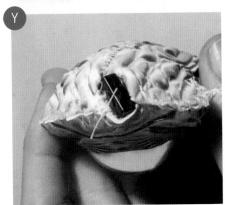

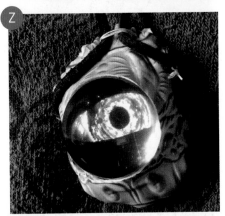

Find this project online at learn.adafruit.com/eye-of-newt.

the back of the OLED display as shown in Figure P. Carefully trim and solder all the remaining wires to the OLED display.

Plug your battery in and slide it between the OLED display and the rest of the components. Wind the wire around and bend the solid core wires until you have a tidy package (Figure Q). Secure everything in place with a few judicious blobs of hot glue.

Flip your switch on and watch your eye dance around (Figure R). Cover the photocell to watch the pupil dilate!

7. MAKE THE CASE

Cut a small piece of fabric about 8"× 8". Place your cabochon in the center and trace around it on the wrong side of the fabric. Cut a hole for the cabochon that's a little smaller than your mark so the cabochon won't fall through (Figure S).

Fold your fabric in half around your electronic eye and mark where it meets itself. Sew the raw edges together with the right sides facing inward (Figure T). Flatten the fabric so the hole is on top and the seam is at the center back. Stitch a curved edge about 1" below the hole (Figure U). Make sure the electronics fit nicely inside.

Place your cabochon into the hole, face down (so you're looking at the flat side). Run a bead of glue all around the edges to hold it securely in place (Figure V). Turn your case right side out and gently slide the electronics inside with the switch and photocell coming out the open top. With a utility knife, make a small slit above the USB port (Figure W). Make another hole for the photocell sensor to poke through (Figure X). Sew up the top of the case with a needle and thread, leaving the on/off switch accessible (Figure Y). I colored my on/off switch with a paint pen so it blends in better with my case.

Finish up by attaching a necklace cord (Figure Z), or leave it as-is and keep it safe inside a potion jar (Figure AA). Remember that the OLED screen is really delicate, so don't try and squeeze it into a jar that's a tight fit — you can break the screen if you squeeze it too hard.

TOIL AND TROUBLE!

Charge it up by plugging in a USB cable — the indicator light on the charger will turn green when it's fully charged. Now you're ready to cast your spells. ◗

Easy **Ultrasonic Levitation**

Hack cheap distance sensors to make tiny objects float in mid-air!

Written by Ulrich Schmerold

ULRICH SCHMEROLD lives in Bavaria in the south of Germany and builds devices for people with disabilities. He likes to create projects that get people excited about physics.

Originally published in the German online edition of Make:, *makezine.com/go/acoustic-levitation. Translation by Niq Oltman.*

TIME REQUIRED:
6–8 Hours

DIFFICULTY:
Intermediate

COST:
$35–$50

MATERIALS

» **Ultrasonic distance sensors, HC-SR04 type (2)**
» **Arduino Nano microcontroller board** with USB cable
» **L293D H-bridge IC chip** or a stepper motor driver module using L298N chip
» **Rectifier diodes, 1N4007 (1 or more)**
» **Capacitors: 100nF (1) and 2,200µF (1)**
» **Solderless breadboard** for testing
» **Perf board, about 2"×2"** for permanent circuit
» **Styrofoam beads (expanded polystyrene)** for levitating
» **Jumper wires and/or hookup wire**
» **Power supply, 9V–12V** A controllable bench power supply is great, or a wall wart. You can even try a 9V battery.

OPTIONAL, FOR HOUSING:

» **Copper pipe, 0.7" (18mm), about 6" long, with fittings: tee (1) and 90° elbows (3)**
» **Scrap wood** for base

TOOLS

» **Soldering iron and solder**
» **Wire cutters / strippers**
» **Computer with Arduino IDE** free download at arduino.cc/downloads
» **Oscilloscope, two-channel (optional)**
» **Hacksaw or pipe cutter (optional)** for copper pipe
» **Woodworking tools (optional)** for base

Ulrich Schmerold

If you'd like to experiment with ultrasonic levitation — making objects hover in mid-air using just the energy in sound waves! — you don't need any scientific equipment, complicated control loop setups, or expensive kits. An Arduino, a stepper motor driver, and a repurposed distance sensor will do.

Yeah, we admit it, our Micro Ultrasonic Levitator won't levitate any heavy items. But it's fascinating enough to watch tiny styrofoam balls hover like magic.

In contrast to magnetic levitation, the ultrasound method does not require a control loop to stabilize the hovering object. Using acoustic levitation, the object simply lodges itself into one of the nodes of a standing acoustic wave. And you can make multiple items hover on top of one another simultaneously, evenly spaced apart!

In our 2/18 issue of *Make:* German edition, we published two alternative approaches to ultrasonic levitation devices, one of them based on a transducer taken from a gutted ultrasonic cleaner.

But a third approach, the one we'll show you here, based on an inexpensive distance sensor, is by far the easiest.

1. DISASSEMBLE THE ULTRASONIC SENSOR

This project is based on ultrasonic transducers of the kind used in distance sensors such as the HC-SR04 module, which can be sourced from eBay for less than $2 (Figure A).

These modules contain one transducer operating as a transmitter (T), and another serving as a receiver (R). In principle, the T transducer is the better pick for use as an actual transmitter, so we bought two sensors and pulled the T transducers off each of them. (In a pinch, you could buy just one sensor — the R transducer also transmits well enough for your first experiments.)

Desolder both transmitter transducers. While you're at it, disassemble one of the receiver transducers (Figure B). Please don't throw away the small, wire screen from the receiver— it turns out to be unexpectedly useful.

The transducers are designed to operate at 40kHz, the frequency at which they work most efficiently. That signal will be generated by your Arduino Nano.

2. UPLOAD THE ARDUINO CODE

The Arduino sketch below performs most of its work in the **setup()** stage. First, it sets all analog ports to be outputs. Then, Timer1 is configured to trigger a *compare interrupt* clocked at 80kHz. Each interrupt simply inverts the state of the analog ports. This turns an 80kHz square signal into a full wave cycling at 40kHz. There's nothing left to do for the **loop()** part of the code!

```
byte TP = 0b10101010; // Every
other port receives the inverted
signal

void setup() {
  DDRC = 0b11111111; // Set all
analog ports to be outputs
  // Initialize Timer1
  noInterrupts(); // Disable
interrupts
  TCCR1A = 0;
  TCCR1B = 0;
  TCNT1 = 0;
  OCR1A = 200; // Set compare
register (16MHz / 200 = 80kHz
square wave -> 40kHz full wave)
  TCCR1B |= (1 << WGM12); // CTC
mode
  TCCR1B |= (1 << CS10); // Set
prescaler to 1 ==> no prescaling
  TIMSK1 |= (1 << OCIE1A); //
Enable compare timer interrupt
  interrupts(); // Enable
interrupts
}
ISR(TIMER1_COMPA_vect) {
  PORTC = TP; // Send the value of
TP to the outputs
  TP = ~TP; // Invert TP for the
next run
}

void loop() {
  // Nothing left to do here :)
}
```

The full code and schematics are available for free download as a ZIP archive at makezine.com/go/micro-ultrasonic-levitator.

3. BUILD THE CIRCUIT

In theory, you could hook up both transmitters directly to the analog ports on the Arduino Nano, as they require very little

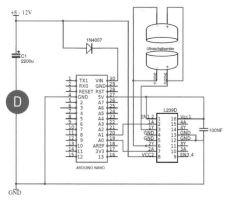

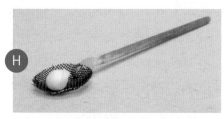

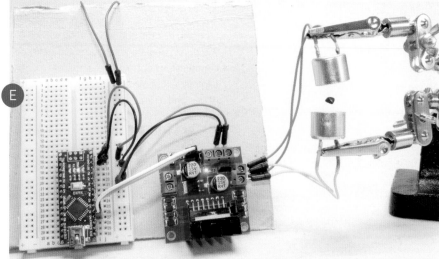

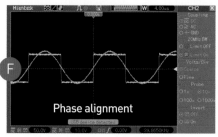

Phase alignment

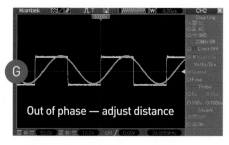

Out of phase — adjust distance

current. However, this would limit you to the 5 volts supplied by the Arduino, greatly reducing the power to levitate. To boost the signal, you'll use an L293D type H-bridge chip, the kind used in some stepper motor drivers. If you're anxious about working with the bare L293D IC directly, you can substitute an L298N-type stepper driver board (Figure C). Just connect two of the four inputs with the Arduino's ports A0 and A1, and wire up GND and 5V as shown in the schematic (Figure D).

If you're constructing the circuit using the bare IC on a perf board, make sure to include the two capacitors. These will filter out the line noise caused by the transducers, which is likely to "blow out" the Arduino, forcing it to keep rebooting.

4. MAKE THINGS LEVITATE!

Start by positioning the transmitters about 20mm (0.8") apart, using a helping-hand tool or something similar. You'll find the exact distance by trial and error. Figure E shows our completed prototype setup, with the stepper module, breadboard, and helping hand.

The distance must be exactly right to create a standing wave with sufficiently strong regions of high and low air pressure.

You can estimate the distance using the following formula, based on the speed of sound at room temperature, 343m/sec (1,125ft/sec):

343,000mm/sec / 40,000Hz = 8.575mm

So, you'd expect to find standing waves at 8.575mm (0.338") or a multiple of that value. But the distance between the transmitter screens is not the same as the area enclosed by the sound wave, so the result won't be quite right. You'll end up moving things around slightly until you get it working.

A two-channel oscilloscope, if available, can help you find the right distance. Connect one channel to the Arduino, and the other to one of the two transmitters (be sure to disconnect it from the board for this measurement). When the distance is just right, the sine wave from the ultrasound receiver should be exactly in phase with the square wave signal from the Arduino (Figures F and G).

Remember that wire screen you saved from the ultrasound receiver? Glued to a toothpick (Figure H), it will help you put those little styrofoam balls in place because it's acoustically transparent. (You'll probably

struggle if you try to use your hands, or tweezers, instead. They'll deflect or perturb the sound from the transducers such that a standing wave may not form at all, or will be too unstable.)

Until you get your first objects to hover, you'll need a bit of patience:

» If it appears as though the balls want to start hovering, but then fall down, try using smaller pieces of styrofoam. They don't need to be round, either. In fact, we found that irregular-shaped pieces appeared to hover more easily.

» Are your hovering objects dancing around wildly? Try reducing the supply voltage. You can use additional 1N4007 diodes wired in series to do this. Each diode will reduce the voltage by around 0.7 volts. With our 12V supply, we got the best results at somewhere between 9V and 11V. (It's probably easiest to just use a variable-voltage bench supply if you can.)

» Once the first polystyrene object is hovering, you can try placing additional objects into the other nodes of the standing wave (Figure I). You'll be impressed!

5. MAKE IT PRETTY (OPTIONAL)

When you're ready to mount your ultrasonic levitator permanently, you can put it in a nice-looking housing to create a coveted desktop toy.

Our final version is made from 18mm (0.7") copper pipe bought at a hardware store (Figure J). We designed the distance between the two transmitters to be exactly 37mm (about 1½") from screen to screen, a number that we worked out experimentally. Your millimeters may vary.

The control board fits easily into the socket in the base (Figure K), provided that you build the final circuit using the L293D IC. On the right, you can see the jack for our 12V power supply.

Happy levitating! ⊘

[+] Watch the Micro Ultrasonic Levitator in action on the project page at makezine. com/projects/micro-ultrasonic-levitator, and see a cool Harvard levitation demo using Schlieren photography to visualize the acoustic standing waves at youtu.be/ XpNbyfxxkWE.

SCALING UP:
Ultrasonic Phased Arrays

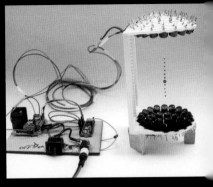

The **Ultraino** open source project takes a similar approach to ultrasonic levitation but it's more powerful. Led by **Dr. Asier Marzo** at Bristol University, this project uses an Arduino Mega and custom amplifier shield to control **phased arrays** of 64 transducers packed into a 3D-printed case. It's capable of levitating liquids, computer chips, and insects, among other things. You can find a detailed how-to at instructables.com/id/ Ultrasonic-Array.

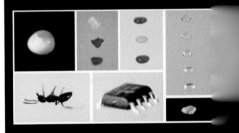

A scientific paper on the Ultraino project describes other fascinating uses for ultrasonic phased arrays, including **steerable mid-air haptic feedback**, **wireless power transfer, parametric audio loudspeakers**, and even **acoustic tractor beams**. Read it free online at ieeexplore.ieee.org/document/8094247.

Another interesting levitation paper describes "**TinyLev**: A multi-emitter single-axis acoustic levitator" (doi. org/10.1063/1.4989995), Marzo's earlier project using 72 transducers. That build is online at instructables. com/id/Acoustic-Levitator; at the end, a "**MiniLev**" variant using only two transducers is described, much like the project shown here. You'll also find a great demonstration video and interview with Marzo, by **Dianna Cowern** aka **Physics Girl**.

Written and photographed by Bob Knetzger

Hot Mods

Cut, shape, and customize Hot Wheels cars to suit your style — and **win prizes**!

TIME REQUIRED:
3–4 Hours

DIFFICULTY:
Easy

COST:
$10–$20

MATERIALS
» **Acetone or nail polish remover**
» **Sugru**
» **Bits of plastic rod, styrene sheet, etc.**
» **MEK solvent**
» **Enamel paints**
» **Cyanoacrylate (CA) glue** aka super glue
» **Laser printer decal sheet**
» **Soft towel**
» **2-56 screws (optional)**

TOOLS
» **Small paintbrushes**
» **Center punch**
» **Drill and bits** including a large center drill bit with shallow drill angle
» **Jeweler's saw**
» **High-speed rotary tool** e.g. Dremel, with grinding and sanding bits
» **Candle**
» **Laser printer**
» **2-56 tap (optional)**

2018 is the 50th anniversary of the introduction of Hot Wheels, Mattel's line of die-cast toy cars. You can celebrate in true maker style by creating your own customized Hot Wheels car.

There's even a contest for *Make:* readers. Submit photos of your custom Hot Wheels by Nov. 30, 2018, and official Mattel Hot Wheels designers will judge the winners and award prizes. Go online to makezine. com/go/hot-mods-contest to enter!

1968 was a huge year for the toy business, with many best-selling (and still well-known) toys like Easy-Bake Oven, G.I. Joe, and games like Operation and Twister. But no toy is bigger than Hot Wheels in total number sold — over 4 billion to date!

Back then the world of die-cast cars was pretty tame, with realistic scale models of sedate sedans, dumpy trucks, and a few race cars. British toy company Lesney sold their Matchbox 1/64 scale vehicles in a "matchbox" package; very cute, and focused on collecting. Mattel revved up the toy car business with their revolutionary new approach, stressing bold new designs and performance.

The original Hot Wheels cars had special low-friction Delrin bearings in mag styled wheels. Instead of straight pieces of stiff wire, Hot Wheels had flexible, thin wire axles formed to act as torsion bars, giving the tiny cars a functioning four-wheel, independent suspension (Figure Ⓐ). To really show off the speed and performance, Mattel engineers also created the iconic flexible orange track along with motorized boosters, high-speed banked turns, jumps, and loop-de-loops.

Just as revolutionary as the technical designs were the aesthetics. The cars were painted with wild "Spectraflame" paint, in transparent candy colors that let the die-cast metal gleam and sparkle. The original car designs included Camaros, Firebirds, Mustangs, and famous custom show cars like the Silhouette, Deora, and Ed "Big Daddy" Roth's Beatnik Bandit (Figure Ⓑ). When the first prototype car was shown to Mattel founder Elliot Handler, he said: "That's one set of *hot* wheels you've got there!" The name stuck.

Mattel also revolutionized the marketing of die-cast cars with exciting promotions.

Mattel

BOB KNETZGER
is a designer/inventor/
musician whose award-
winning toys have been
featured on *The Tonight
Show*, *Nightline*, and
Good Morning America.
He is the author of *Make: Fun!*, available
at makershed.com and fine bookstores
everywhere.

Kids were thrilled to re-create the exciting drag strip battles between real-life rivals Don "The Snake" Prudhomme and Tom "The Mongoose" McEwen on their bedroom floor, complete with popping drag 'chutes!

Today, the original sixteen Hot Wheels cars (called "Redlines" for their red sidewall tires) are treasured by collectors. The holy grail of vintage Hot Wheels cars is the limited-run Beach Bomb VW surfer van prototype with rear-loading surfboards. One famously sold for over $100,000 and was featured on PBS's *Antiques Roadshow*.

CUSTOMIZE YOUR HOT WHEELS CAR

Ready to mod your 'rod? Here's an example of an easy project to help get you started. My toy invention business partner, Rick Gurolnick, races in this 1960 Porsche 356 Speedster (Figure C). Its number 60 and "Doctor Dreadful" livery are well known on the vintage racing circuit. I wanted to make a tribute Hot Wheels version of his car, and fortunately Mattel produced a vintage Porsche that I could use as a starting point (Figure D). It was a coupe, so I needed to

remove the roof, cut down the windshield, scratch-build the roll bar, as well as paint and create custom decals.

1. Remove the packaging

I wanted to be able to put the finished car back into the original blister card, so I needed to remove the car without damaging the packaging. How? Brush acetone (nail polish remover) on the *backside* of the card along the bottom and sides where the vacuum-formed blister is glued on (Figure E). Apply liberally and let it soak in, right through the card. The acetone will gently dissolve the glue without affecting the printing. I left the top edge glued on so I could replace the car later (Figure F).

2. Drill out rivets

Hot Wheels cars are fastened together by "heading" the posts of the metal body, mushrooming them like a small rivet. To disassemble, place the car on a soft towel and carefully drill out the rivet heads (Figure G). Start by punching a divot with a center punch, then use a large drill with a shallow drill angle so that you remove just enough

of the top of the post to release the car body (Figure H). That will preserve most of the post for reassembly. Use a center drill instead of a twist drill for more control and minimum material removal.

3. Disassemble

Once drilled out, disassemble the car parts. This car has a separate interior, body, chassis, and plastic windshield (Figure I).

4. Cut and shape

To remove the car's roof I used a jeweler's saw (Figure J). Hot Wheels cars are made of *zamac*, an alloy of zinc, aluminum, magnesium, and copper. It's soft enough that you can cut, grind, drill, and sand to make the modifications of the metal body (Figure K, following page). I also cut the plastic windshield down as shown in the next step.

5. Create your own custom parts

The real race car has a tonneau cover, so I used some Sugru to fill in the passenger side of the interior and build up enough to make a cover (Figure L, following page). The black Sugru has a nice matte finish that's just right.

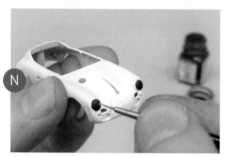

I hand-fabricated a roll bar from pieces of 1/16" plastic rod. I used a candle flame to carefully heat and soften the rod just enough to bend, then held it at the needed angle until cooled and set. I used MEK to solvent-bond the rod pieces together to make the roll bar and supports, then painted it with flat black paint (Figure M).

6. Paint

The real race car's basic body color is white, so I sprayed some primer and solid white enamel paint on the body. I brushed on a few painted details with black, silver, and red enamel paint (Figures N and O).

I also fabricated a tiny hood handle from bent styrene sheet, painted it silver, and super-glued it in place.

7. Reassemble

Press the body posts back into the chassis holes. Often there's still enough friction and material in the posts to hold the car together gently for display. For a more rugged assembly, drill and tap holes in the post and reassemble with 2-56 screws.

8. Do decals

To make the decals, I found images online of the Mobil Pegasus and other logos, and created text images for the numbers and driver name in a matching font. You have to print these on a special laser printer decal sheet, because inkjet inks aren't waterproof.

Trim the decals, loosen in water, and slide into place on the car — done (Figure P). Looks just like the real car!

Here's the final result, back in the original package and with a collector protective clamshell, ready to display (Figure Q). And you can still open up the package for play.

REV UP YOUR RIDE

Now it's your turn! Which cool custom Hot Wheels car will you make? A gleaming metal flake show car? "Rat rod" street racer? Lowrider with custom spoke wheels? Post-apocalyptic monster truck? It's up to you!

And Happy 50th Birthday, Hot Wheels! ⊘

Enter the Hot Mods contest by Nov. 30:
makezine.com/go/hot-mods-contest

Hot Wheels history:
hotwheels.mattel.com/explore/HW_50th

Parts and paints: redlineshop.com

Expert mods, restorations, and customs:
YouTube channel BaremetalHW

1+2+3 Follow Me Eyes

Written by Eduardo Talbert

Here's a super easy and fast project: eyes that seem to follow you wherever you stand! Use these for your Halloween props, spell books, or practically anything. They make awesome refrigerator magnets.

1. IRISES

Find images of eyes online, then print (Figure A) and cut out irises about ½" in diameter. I like to use Terra's Halloween Eyes, makezine.com/go/terra-eyes.

Brush glue on the backs of the gems (Figure B), apply the irises facedown, and press out any air bubbles (Figure C). Let dry, then add a second coat to "decoupage" the iris to the gem.

2. CAPILLARIES

Pinch a few red yarn fibers (Figure D) and arrange them on the back of the gem (Figure E). It's OK if they stick out the sides. Coat with glue.

3. SCLERA

While the glue is wet, cover the back of the eye with white paper, press out bubbles, and let dry (Figure F). Cut off excess yarn and paper around the edges. For more durable eyes, decoupage once more.

EYE SEE YOU

That's it! You now have eyes that follow you. Place them so they're facing you — on a prop, or with a magnet on the refrigerator. Now walk around and see how the eye seems to be always looking at you! ✪

TIME REQUIRED:
20–30 Minutes

DIFFICULTY:
Easy

COST:
$5–$10

MATERIALS
- » **Clear glass "gems," 1" (2)** aka flat marbles, glass pebbles, etc.
- » **White printer paper**
- » **Red or burgundy yarn scrap, frayed (optional)** for capillaries
- » **Prop or magnets (optional)**

TOOLS
- » **Computer and printer**
- » **Mod Podge** or other clear drying glue
- » **Small paintbrush**
- » **Scissors**

EDUARDO TALBERT is a father of three rambunctious boys, husband to a wonderful wife, and maker of monsters and DIY tutorials at monstertutorials.com.

A

B

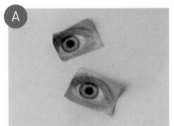
C

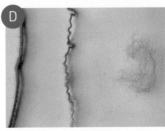
D

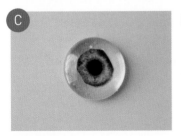
E

F

Holden Johnson

Racing Robots

Inspired by board game RoboRally, open source **RoboRuckus** lets players navigate a zany obstacle course with actual bots

Written by Sam Groveman

My friends and I had always enjoyed the classic board game RoboRally, and the idea to improve it with actual robots was too compelling to pass up. Of course, the actual implementation of our completely open source RoboRuckus proved to be challenging.

THE PREMISE

You, the player, control a robot with the goal of navigating it through the hazardous board and touching a series of up to four checkpoints in order. Each turn, the players

are dealt cards that contain movement instructions (move forward, turn left, back up, etc.), from which they pick five to be their robot's instructions for that turn (Figure Ⓐ). Once chosen, the instructions cannot be changed. After all players have selected their robot programs, all the robots move at once, pushing and jostling each other in their dash to reach the checkpoints.

NAVIGATION

Optical line following is the simplest and most popular method, but that requires having high-contrast lines all over the board. The solution we eventually came up with was to use magnetic line following, with magnetic mounting tape and Hall effect sensors to mimic the photosensors used in optical line following. These sensors were later replaced with more reliable magnetometers.

ROBOTS

With two primary principles of simplicity and inexpensiveness in mind, I decided to go with a AA battery case for the robot body and continuous rotation servos for movement. A seven-segment LED display and a piezo buzzer would provide user feedback, and the robot would connect to the control server via Wi-Fi.

Our first prototype (Figure Ⓑ), controlled by an Arduino Pro Mini, proved the Hall effect sensors too insensitive to follow the magnetic tape lines on the other side of a piece of cardboard. The sensors were replaced with digital compasses — one in front and one in back, in order for the bot to be able to drive forward and back up. This required two I2C buses, so we switched from the Arduino Pro Mini to a Teensy LC.

In our second prototype (Figure Ⓒ), we

SAM GROVEMAN
Although Sam obtained his Ph.D. in chemistry, he's always loved using computers and electronics, especially when making, or attempting to make, fun and interesting projects.

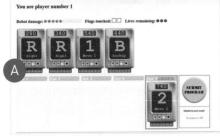

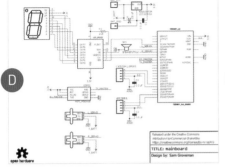

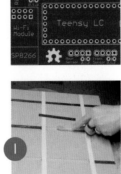

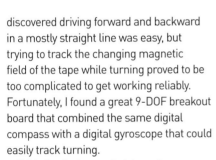

discovered driving forward and backward in a mostly straight line was easy, but trying to track the changing magnetic field of the tape while turning proved to be too complicated to get working reliably. Fortunately, I found a great 9-DOF breakout board that combined the same digital compass with a digital gyroscope that could easily track turning.

With the design settled, I used some schematics (Figure D) that I whipped up in EAGLE (available at github.com/ ShVerni/RoboRuckus) to order custom PCBs from OSH Park (Figure E) to speed the manufacturing process, and to look better. We assembled an army of robots, with a little flair (Figure F).

GAME BOARD

Since the robots turned out to be around 4" at their largest dimension, I decided the board size should consist of a 12×12 grid of five-inch squares laid out on 1/8" plywood. We placed a grid of magnetic tape (Figure G), and then further divided the board into nine sections of 4×4 squares to make assembling and transporting it easier (Figure H).

I recreated a classic RoboRally board from scratch (with some modifications) so that it would be high resolution enough to be printed — on vinyl to help smooth out bumps between joints — in the size needed. As we continued to test and fine-tune the robots, we discovered the board itself had warped, and that the magnetic tape running underneath the plywood proved too indistinct to track reliably, especially at the intersections where two pieces overlapped.

We ended up doing a complete redesign, and decided to go with a thicker, sturdier 1/2" MDF. We set the tape into the top, bringing it closer to the sensors, and embedded a rare earth (neodymium) magnet at each intersection to ensure that the magnetic field there was significantly higher than the lines. When all the intersections had their magnets, the magnetic strips could be installed (Figure I). These changes significantly improved the performance of the robots. As a bonus, the original vinyl mat could be reused.

THE CODE

The rules of RoboRally are relatively simple and algorithmic, so they actually translated pretty well to computer code. The game server itself runs on a Raspberry Pi, which broadcasts its own Wi-Fi network. ASP.NET Core running on the Pi provides the web interface to which all the players and robots connect. The game server then coordinates the game play, robot positions, and player inputs.

BUILD IT

While we're mostly happy with the game, there are still some refinements we'd like to make. Chief among them is a 3D-printed chassis for the robots. Also, we'd like to redesign the power supply to provide a more consistent voltage to the servomotors, possibly using LiPoly batteries.

It took us a little under a year of sporadic monthly tinkering to get this project to its first working state, but now that we have all the major speed bumps worked out I believe a small group could build their own version in a weekend of solid effort. ◉

[+] Read more of the RoboRuckus story and find detailed instructions at makershare.com/projects/roboruckus and roboruckus.com.

[+] Find this issue's other Mission to Make: favorites in "Show & Tell" on page 80 and online at makershare. com/missions/mission-make-vol-65.

Write On Time

Written by Tucker Shannon

Build a cute robot arm to draw the current time in luminescent numerals

A

TUCKER SHANNON is a mechanical/software engineer from Bend, Oregon. He loves building fun creations that combine art, tinkering, and engineering.

Tucker Shannon

My Glow-in-the-Dark Plot Clock uses an Arduino Uno, two 9g servomotors, a UV LED, and a phosphorescent material to "draw" the time on demand. Whenever the button is pressed, a robot arm writes the time on the screen in glowing numerals that shine for several minutes before slowly fading away.

This project was inspired by the amazing whiteboard plot clock by Johannes Heberlein in Nuremberg, Germany (thingiverse.com/thing:248009), which writes the time with a dry-erase marker, and then erases it!

I really liked that clock, but I wanted a writing system that would last longer than the ink in a whiteboard marker. I decided to use a glow-in-the-dark material instead of the whiteboard, because it produces a

visually appealing glow effect, "erases" itself, and can't run out of ink. This glow effect, along with the 3D-printed enclosure, ensures that your plot clock will work reliably for years to come.

My previous iteration of this plot clock used a more expensive UV laser. Here I have replaced the UV laser with a UV LED that produces the same effect at a fraction of the cost. In fact, all the components required to build this project can be purchased for around $8 from AliExpress — that's less than the laser cost alone. But shipping takes a long time. If you're worried about quality and shipping time, Amazon and an official Arduino will work better.

You can download the 3D print files and Arduino code from thingiverse.com/thing:2833916, and find all the AliExpress

TIME REQUIRED:
1–3 Hours

DIFFICULTY:
Intermediate

COST:
$15–$50

MATERIALS
» **Arduino Uno microcontroller board** with USB cable
» **Ultraviolet (UV) LED, 5mm**
» **Micro servomotors, 9g (2)** such as AliExpress #32839396457
» **Real-time clock (RTC) module** DS1307 type, I2C input/output. I used the Tiny RTC board, AliExpress #1086254258. Upgrade: the DS3231 keeps better time.
» **Pushbutton switch, momentary, normally open, 12mm**
» **Hookup wire** 22 gauge or similar
» **Rubber feet (4)**
» **Glow-in-the-dark sticker sheet** AliExpress #32801864131
» **Machine screws, hex, M3×8mm (10)**
» **3D printed or laser-cut parts** Download the free files for the arm and enclosure at thingiverse.com/thing:2833916 or thingiverse.com/thing:2845462.

TOOLS
» **Soldering iron**
» **3D printer or laser cutter**
» **Computer with Arduino IDE** free download at arduino.cc/downloads

links there too. I also made a laser-cut version (Figure A) if you've got laser cutter access: thingiverse.com/thing:2845462.

BUILD YOUR GLOWING PLOT CLOCK
I created a video assembly guide at youtu.be/-MnolVyKqvo to show you how it all goes together. Here's a quick step-by-step:

1. Print (or cut) parts
Gather all the components listed, and 3D print (or laser cut) the enclosure and arm files found on Thingiverse. Print using PLA or ABS, without supports. I used PLA but most materials will work.

2. Wire the circuit
Solder 2 wires to the momentary pushbutton terminals, and 2 to the 5mm UV LED. Insert the servomotors into the front of the housing, and thread the LED wires through the front too.

Mount the Arduino Uno inside the housing using M3 screws. Solder wires onto

the real-time clock (RTC) module as shown in the schematic diagram (Figure B), then mount the RTC in the housing using M3 screws.

Now connect the wires from the RTC, servos, button, and LED to the Arduino, carefully following the schematic.

3. Assemble the enclosure and arm
Attach the front, top, and bottom of the enclosure using M3 screws (Figure C). Then connect the arms together using M3 screws.

Insert the LED into the arm assembly, then attach the focusing cone over the LED.

Attach servo horns to the arm assembly and finally, attach the arm assembly to the enclosure (Figure D).

Cut the glow sticker to size and apply to the front of the enclosure. Add rubber feet to the bottom.

4. Calibrate and upload code
Plug your Arduino into your computer. Open the Arduino IDE, then follow my code and calibration guide video at youtu.be/4viW9ADqX2w. You'll install the Time and DS1307RTC libraries by Michael Margolis, set the time on the RTC, then calibrate the project code file *Arduino_Code_Glow_Plot_Clock.ino* (Figure E) so your servos are drawing a perfect rectangle around the glow sticker. That way they'll always draw the time in the right place.

My code is a modification of the code in the whiteboard plot clock. I changed a few things to better suit the new glow design:
» Use of 2 servos instead of 3
» Ability to toggle 12-hour or 24-hour times
» Pushbutton, instead of write every minute
» Different servo calibration methods
» Turns on an LED instead of lowering a marker to the surface

QUALITY TIME
Now whenever you push the button, your plot clock's robot arm will write the hour and minute in glowing green. It looks awesome in the dark (Figure F) and it's easily seen during the day too (Figure G).

Please share your builds and comments at makershare.com/projects/glow-dark-uv-plot-clock-diy. And check out my YouTube channel TucksProjects for a new laser version that can also do the weather report, talk to Google, and write messages. ⊘

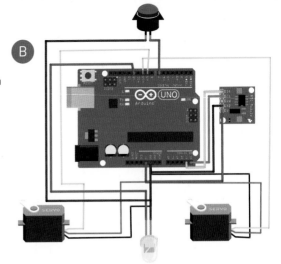
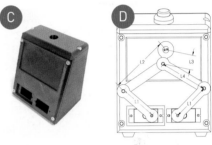

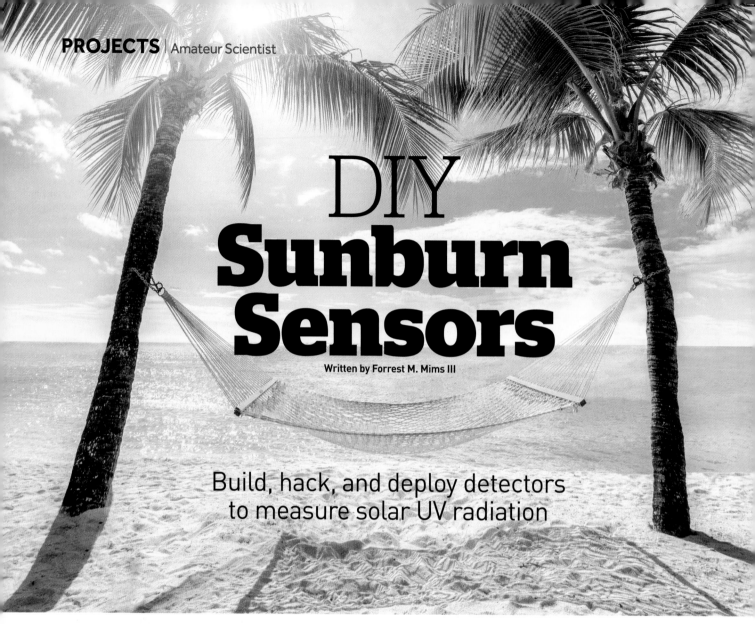

DIY
Sunburn
Sensors

Written by Forrest M. Mims III

Build, hack, and deploy detectors to measure solar UV radiation

UV INDEX
(at local noon under clear skies)

11+	Extreme
10	
9	Very high
8	
7	High
6	
5	
4	Moderate
3	
2	Low
1	

Sunlight is essential for our survival, but too much of a good thing can sometimes be harmful. That certainly applies to the invisible ultraviolet (UV) wavelengths of sunlight, for too much exposure to UV can cause sunburn and even skin cancer. Yet, moderate exposure is important, for both mammals and reptiles depend on ultraviolet sunlight to manufacture the vitamin D that supports the growth of bones and fights some diseases.

ULTRAVIOLET AND HEALTH

Light is specified according to its wavelength. For example, green light near the peak response of human vision has a wavelength of 500 nanometers (nm) or half a micrometer. If UV wavelengths were visible, they would appear adjacent to the violet portion of a rainbow. The UV wavelengths are divided into three categories, each having distinctive effects on plants and animals:

» **UVA: 320 to 400nm.** UVA wavelengths penetrate deeper into skin than UVB and UVC. Excessive UVA exposure can lead to skin wrinkling. Recent research suggests that excessive UVA can also lead to skin cancer.

» **UVB: 280 to 320nm.** Most UVB is absorbed by the ozone layer, but some leaks through. UVB causes erythema, the reddening of the skin that precedes sunburn. Excessive UVB exposure can lead to skin cancer. UVB can also damage eyes.

» **UVC: 100 to 280 nm.** UVC rapidly kills viruses and bacteria. It is absorbed in the dead cells in the uppermost layer of human skin, where it does not lead to erythema. UVC exposure to living skin cells can cause erythema. UVC can also cause eye damage. Fortunately, UVC is totally blocked by the ozone layer.

It's best to avoid more than several minutes of UVB exposure whenever your shadow is shorter than your height. You can minimize your exposure by applying sunscreen and wearing a brimmed hat and a long-sleeved shirt. You can learn more about UV hazards at epa.gov/sunsafety.

Total protection from UV isn't necessary for most people, for UV stimulates the production of vitamin D in the skin. Vitamin D allows the body to metabolize calcium and provides protection from rickets in children and osteoporosis in the elderly. It may also protect against several kinds of internal cancer. Learn more at vitamindcouncil.org.

THE UV INDEX

The erythemal wavelengths are mainly between 295 to 320nm in the UVB but also include some UVA out to 370nm. This range of wavelengths is known as the *erythema action spectrum*, and its intensity is specified by the UV Index (UVI):

» Low: 0–2
» Moderate: 3–5
» High: 6–7
» Very High: 8–10
» Extreme: 11+

The peak UVI occurs during the summer months. It can exceed 12 where I live in Texas and 20 at Hawaii's Mauna Loa Observatory. NOAA and the EPA provide forecasts of the UVI online at epa.gov/sunsafety/uv-index-1.

BUILD A SIMPLE UVA-UVB METER

Aluminum gallium nitride (AlGaN) photodiodes greatly simplify UV measurements, since expensive filters are not required. Several UV sensors that use AlGaN photodiodes are available. One is

Adafruit's GUVA-S12SD Analog UV Light Sensor (Figure A), which uses a GenUV GUVA-S12SD GaN photodiode from Roithner LaserTechnik that responds from 240 to 370nm. While this diode is widely used to measure the UV Index, it's not ideal since it's more sensitive to UVA than UVB.

Figure B shows the wiring diagram for a simple UV meter that uses the Adafruit sensor. The readout is a miniature 3½ -digit LCD voltmeter, the Lascar EMV 1025S-01. Its connection wires emerge from a hollow, threaded stud that allows the meter to be attached to a flat surface.

The readout is powered by 6 volts from a pair of 3V lithium coin cells installed in a Mini Skater dual CR2032 coin cell holder. The holder is modified to also provide +3V for the sensor by twisting ¾" of exposed wrapping wire around one end of the common contact that links both cells inside the holder. Inserting the cells and closing the holder will lock this 3-volt wire in place.

The components can be installed neatly in a 3"×1½"×¾" Altoids Arctic mint tin (Figure C).The UV sensor fits inside the recessed lid of the Altoids tin, where it can be secured with a single ¼" or ½" 6-32 screw and nut. Or it can be protected by being mounted inside the tin, as I did. This requires boring two holes in the upper side of the tin: a ⅛" hole 9⁄16" from the opening for the UV photodiode, and an adjacent 3⁄32" hole 7⁄16" from the opening for a 2-56 screw and nut to hold the sensor board in place (Figure D).

To provide a better full-sky response, insert two layers of Teflon film between the photodiode and its hole in the box. Teflon must be used, since most other diffusing materials will not transmit UV.

TIME REQUIRED:
2–3 Hours

DIFFICULTY:
Intermediate

COST:
$15–$75

MATERIALS
» **Analog UV Sensor Board with GUVA-S12SD sensor** Adafruit #1918, adafruit.com
» **Dual CR2032 coin cell holder with switch** Amazon #B074FZYCKP
» **Switches, miniature DPDT toggle (2)**
» **Panel voltmeter LCD display** Lascar EMV 1025S-01, lascarelectronics.com
» **Altoids Arctic mint tin**
» **Wrapping wire** such as 28 or 30 AWG wire gauge
» **PVC pipe, ¼" ID, 17⁄32" OD, 11⁄16" long**
» **2-56 screw and nut**
» **Teflon film** from a sewing supply store
» **Teflon disc, ½" diameter, 0.4mm thick** I used Cox #49DDISC, coxengines.ca.

OPTIONAL:
» **Data logger, Onset 4-channel, 16-bit** onsetcomp.com
» **Miniature stereo phone plug**
» **UVB photodiode, SMD package** Roithner #GUVB-S11SD, roithner-laser.com/pd_uv.html
» **Resistor, 10MΩ, SMD package**

TOOLS
» **Voltmeter**
» **Wire wrap tool**
» **Wire cutters**
» **High-speed rotary tool** e.g., Dremel
» **Soldering iron and solder** preferably a low-power USB iron

eyetronic - Adobe Stock, Forrest M. Mims III

FORREST M. MIMS III (forrestmims.org), an amateur scientist and Rolex Award winner, was named by *Discover* magazine as one of the "50 Best Brains in Science." His books have sold more than 7 million copies.

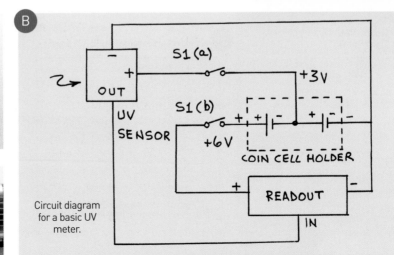

Circuit diagram for a basic UV meter.

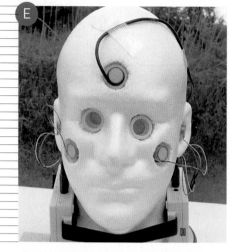

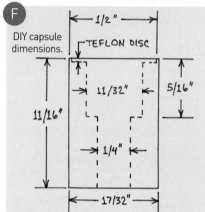

DIY capsule dimensions.

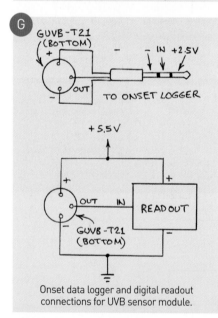

Onset data logger and digital readout connections for UVB sensor module.

Forrest M. Mims III

A UVB METER HACK

You can greatly improve the Adafruit UV sensor's response to the erythemal action spectrum for a UVI of 3 and above by swapping in a GUVB-S11SD UVB photodiode and increasing the feedback resistor to 10 megohms. While this hack works well for a UVI of 3 and above, low values of UV will give a false reading.

Begin by placing a white washcloth on your workbench to trap the tiny chips should you drop them. Then use a soldering iron with a very small tip, such as a Mega Power USB-powered iron, to carefully remove the original photodiode and feedback resistor just above it. My technique is to melt the solder on one side of the photodiode and gently tilt it upward. I then melt the solder on the second side. This leaves space to use solder wick to capture the solder holding the resistor in place.

Next, use masking tape to secure one side of the new photodiode in place. The notch in the chip's corner must face toward the S in the word *Sensor*. Carefully heat the exposed junction of the chip and the board to melt solder between the two. If necessary, apply a bit of very thin (0.02") solder. Remove the tape and solder the second side of the chip. Repeat this procedure to replace the original feedback resistor with a 10-megohm substitute.

I've done this with three Adafruit UV sensors. One failed and two worked well for a UV Index of 3 and above. But I don't recommend this hack unless you have prior experience removing and installing tiny surface-mount chips.

A HIGH-QUALITY UVB SENSOR

My best results with DIY UV Index meters have been with the GenUV GUVB-T21GH

sensor module from Roithner. This module includes an AlGaN UVB photodiode and amplifier in a TO-5 case with a quartz window. It works so well that I'm using seven of them to measure UVB on a rotating mannequin head (Figure **E**), in a Rolex-sponsored UV survey of the island of Hawaii. The head rotates 1,000 times a day while equipped with various hats and sunglasses.

David Brooks of the Institute for Earth Science Research and Education, an advisor for this project, made holders for the sensor modules that give them an optical response similar to professional-quality UV sensors that cost hundreds of dollars. The DIY holders are machined from ¼" ID PVC and fitted with a ½" disk of 0.4mm-thick Teflon (Figure **F**). The result certainly justifies the $38 (plus shipping) cost of these modules.

The GUVB-T21GH can be connected to the Lascar readout described before and the two can be powered by the same modified 6V coin cell holder. For my research, the GUVB-T21GH is directly connected to an Onset 12-bit or 16-bit analog data logger using wrapping wire soldered to a miniature stereo phone plug that's inserted into the logger, which provides 2.5 volts for the sensor. Figure **G** shows connection details for both methods. Figure **H** shows a GUVB-T21GH module installed in an Altoids tin.

You can use the NOAA/EPA hourly UV Index forecast for your zip code to convert the voltage from the module to the approximate UV Index, as shown in Figure **I**. Use a spreadsheet to make a graph of the data that you can tape to the instrument or keep in a pocket. The high quality of the GUVB-T21GH fitted with a Teflon diffuser is shown in Figure **J**, which compares our DIY sensor with a much more expensive PMA1102 UV detector from Solar Light. ⬤

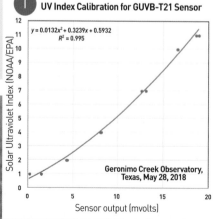

I UV Index Calibration for GUVB-T21 Sensor

$y = 0.0132x^2 + 0.3239x + 0.5932$
$R^2 = 0.995$

Solar Ultraviolet Index (NOAA/EPA)

Sensor output (mvolts)

Geronimo Creek Observatory, Texas, May 28, 2018

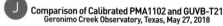
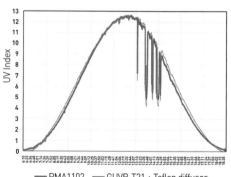

J Comparison of Calibrated PMA1102 and GUVB-T21
Geronimo Creek Observatory, Texas, May 27, 2018

UV Index

— PMA1102 — GUVB-T21 + Teflon diffuser

GIVE A GIFT.
ONE YEAR ONLY $39.99.

Make:

GIFT FROM

NAME _____ (PLEASE PRINT)

ADDRESS/APT. _____

CITY/STATE/ZIP _____

COUNTRY _____

EMAIL ADDRESS (required for order confirmation) _____

☐ Please send me my own subscription of Make: 1 year for $39.99.

GIFT TO

NAME _____ (PLEASE PRINT)

ADDRESS/APT. _____

CITY/STATE/ZIP _____

COUNTRY _____

EMAIL ADDRESS (required for access to digital edition) _____

484GS1

We'll send a card announcing your gift. Make: currently publishes 6 issues annually. Occasional double issues may count as 2 of the annual 6 issues. Allow 4-6 weeks for delivery of your first issue. For Canada, add $9 US funds only. For orders outside the US and Canada, add $15 US funds only. Access your digital edition after receipt of payment at make-digital.com.

BUSINESS REPLY MAIL

FIRST-CLASS MAIL   PERMIT NO. 865   NORTH HOLLYWOOD, CA

POSTAGE WILL BE PAID BY ADDRESSEE

Make:

PO BOX 17046
NORTH HOLLYWOOD CA 91615-9186

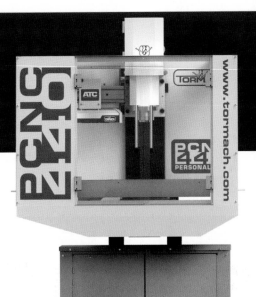

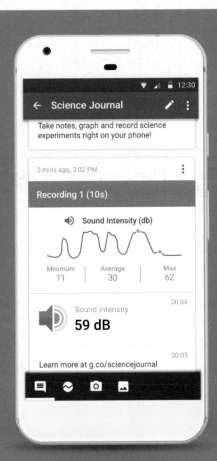

TIME REQUIRED:
1–3 Hours

DIFFICULTY:
Easy

COST:
$5

MATERIALS
» LEDs, 3mm (14)
» Wire, solid core, about 2"–3"
» Cardboard
» Coin cell batteries, 3V (2)
» Jump rings (4)
» Earring hooks (2)

TOOLS
» Soldering iron
» Scissors
» Wire cutters / strippers
» Pliers

CLARE MASON is a techie working in the greater Seattle area. She likes crafts and coding. Preferably at the same time. Find more of her shenanigans on Twitter as @makeandfake.

Flashy Fashion

Light up the room with these lightweight LED earrings Written by Clare Mason

Before attending a fancy event, my friend asked me to create some earrings for her that would light up. I needed to design something that would be lightweight, but I also wanted it to be wearable without the battery for everyday use. I started with a small 3V coin cell battery and built it out from there by playing around with the LEDs until they fit. This is the design I came up with.

Since then, I've built multiple sets of earrings for different people, each time mixing up the colors and sizes of the LEDs. I also use this project to introduce people to electronics, because it requires minimal soldering, creates a simple circuit, and in the end, you get to take home a pair of beautiful earrings.

1. MAKE THE TEMPLATE
Trace your battery onto a piece of cardboard with the approximate thickness of your battery and cut it out (Figures A and B). This will serve as a template to help you arrange your LEDs and hold them in place while you solder.

2. ALIGN LEDS
Slide the cardboard template in between the legs of each LED. Arrange them neatly until they make the shape you want for your earring (Figure C). Make sure to align all the positive legs on one side of the template and negative legs on the other. If you're unsure of the polarity of your LED, you can always check using your battery.

3. SOLDER THE LEDS TOGETHER
Starting on the side with the positive (longer) LED legs, strip and solder a length of wire across the positive legs, as close as you can to the top, connecting them all together (Figure D). Trim the excess wire.

On the side with the negative legs, solder another length of bare wire, this time placing the wire a little over halfway up the legs and soldering it into place (Figure E). Trim the excess wire.

This V shape will now hold the battery in place. Test it by slipping the battery in to make sure everything is working and the LEDs light up. Bend the LED legs if needed to hold the battery more firmly.

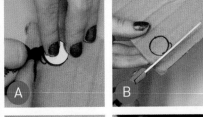

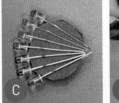

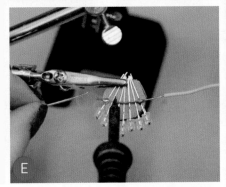

William Lambrecht

4. ATTACH EARRING HOOKS

Solder a jump ring to the top of the positive side (Figure **F**). Next, attach the earring hook by placing a second jump ring into the soldered one, and then attaching the earring hook to the second ring (Figure **G**).

Slip the battery in, and your earring is ready! Complete this process a second time to create two matching earrings. Wear these earrings with or without batteries. You will look great either way!

GOING FURTHER

More or fewer LEDs may be used depending on their size and your personal preference. Feel free to play around with varying sizes and colors to fit your personal style. ⊘

Written and photographed by Meia Matsuda

1+2+3 Inner Tube Earrings
Upcycle bike tubes into elegant adornments

Inner tube rubber, aka "vegan leather," is a durable waste material that is fun to play with and can be remade into jewelry, wallets, belts, and many other items. These earrings are a great conversation starter!

Stop by your local bike shop to pick up punctured inner tubes they're discarding. I used the skinny tubes from road bikes.

1. CUT

Cut two identical parallelograms the length you want your earrings to be. Then cut straight lines across at the same angle, leaving a 1cm spine (or 5mm when flattened). For a polished look, keep all cuts parallel at a uniform width and length.

2. FLIP

Lift the bottom of the first "leaf" over the stem axis to flip it inside out. Leave the second leaf as is. Grab the top of the tube, weave it through the third leaf, and gently pull all the way through so the third leaf is inside out. Repeat, flipping every other leaf.

3. FINISH

Wash the tube and pierce a hole at the top. Push a jump ring through the hole, place the earring hook onto it, and close it up.

You're ready to rock these earrings! Have fun experimenting — with different tubes, earring lengths, and thinner or wider leaves — and getting compliments! ⊘

TIME REQUIRED:
1 Hour

DIFFICULTY:
Easy

COST:
$1–$5

MATERIALS
» Old bicycle inner tube
» Jump rings (2)
» Earring hooks (2)

TOOLS
» Scissors, pushpin, needlenose pliers

1

2

3

MEIA MATSUDA is the founder of Innercycled. She graduated from the University of California, Berkeley with a degree in Sustainable Environmental Design, and she enjoys traveling, design, rethinking waste, and nerding out about city planning.

Written and photographed by Charles Platt

Audible
Aqua

Use an op-amp to build a water purity tester you can hear

TIME REQUIRED:
1–2 Hours

DIFFICULTY:
Easy

COST:
$10–$20

MATERIALS

- » **Breadboard, dual bus**
- » **Patch cords with alligator clips (2)**
- » **9V battery**
- » **Small cups (6)** for liquid samples
- » **Pennies (2)** as electrodes
- » **Plastic or plywood, about ¼"×1"×3"**
- » **Jumper wires, 1" or shorter (10)** for breadboard
- » **Loudspeaker, 2" or 3" diameter, 8Ω**
- » **LM741 op-amp IC chip**
- » **555 timer IC chip**
- » **Resistors: 100Ω (1) and 15kΩ (4)**
- » **Trimmer potentiometer, 100kΩ**
- » **Ceramic capacitor, 0.1µF**
- » **Electrolytic capacitor, 100µF**

TOOLS

- » **Miter saw or handsaw**
- » **Drill and ¼" bit**
- » **Wire strippers**

Can you be electrocuted in the bathtub? Sure — but only because the water's not pure. Many people don't realize that absolutely pure water doesn't conduct electricity. This is because the hydrogen and oxygen atoms don't have free electrons. In a typical domestic water supply, it's the impurities such as sodium, calcium, and magnesium salts that enable electrons to flow.

This interesting fact means that you can assess the purity of water by measuring its electrical resistance. The higher the resistance, the purer the water should be (although this test won't reveal contamination by substances such as organic compounds).

The simple circuit in this project can test water samples without any need for a multimeter.

VOLTAGE DIVIDER

The basic concept is to use a small volume of water in a *voltage divider*. You can then compare the output from the divider with a reference voltage, and amplify the difference. When the output is passed along to a timer chip, you'll hear a sound that changes with the purity of the water.

The concept of a voltage divider is illustrated in Figure **A**. When an input voltage, shown as V_{in}, is applied across two resistors in series, you can calculate the output voltage, V_{out}, at the point between the resistors. If both the resistors are the same value, V_{out} is half of V_{in}. If R1 is twice the value of R2, V_{out} is one-third of V_{in}. And so on. (If you add a significant load to the output, V_{out} will be reduced.)

If you use a water sample instead of R1, V_{out} will vary with the purity of the sample. You can then compare it with a constant reference voltage, shown as V_{ref}, created by a second voltage divider, as in Figure **B**.

OP-AMP

Because the difference between V_{out} and V_{ref} may be small, you need to amplify it. This is easily done with an *operational amplifier*, commonly known as an *op-amp*, such as the venerable LM741. The symbol for an op-amp is shown in Figure **C**, although when you see it in schematics, you'll find that people often don't bother to

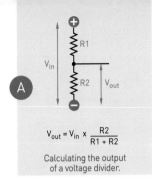

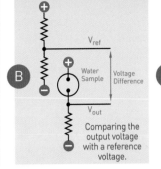

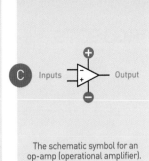

A

$$V_{out} = V_{in} \times \frac{R2}{R1 + R2}$$

Calculating the output of a voltage divider.

B

Comparing the output voltage with a reference voltage.

C

The schematic symbol for an op-amp (operational amplifier).

show the power supply.

The plus and minus symbols inside the triangle identify the "noninverting" and "inverting" inputs, respectively. The output from the op-amp increases when the noninverting input increases, or decreases when the inverting input increases.

Figure D shows the op-amp in a circuit using the water sample and the reference voltage as inputs. I substituted a potentiometer for one of the resistors, so that you can control the amplifying power with negative feedback to the inverting input. (There are many other ways to wire an op-amp, which you can easily find online.) Pin functions of an LM741 chip are shown in Figure E.

I connected the output of the LM741 to a 555 timer, using pin 5 of the timer to control its audible frequency. In the schematic in Figure F, the complete circuit is laid out as you might wire it on a breadboard. You can find a lot more information about timer chips in my book *Make: Electronics*.

TESTING LIQUIDS

Commercial resistive water testing devices use platinum contacts, but that was a bit beyond my budget, so I used a pair of pennies. Try to find some that are as new and bright as possible.

To mount the pennies, you can use a piece of plastic or plywood about 1"× 3" and ¼" thick. Drill a couple of holes ¼" in diameter, about ¾" apart, and make saw cuts across them, as shown in Figure G. Poke your alligator clips through the holes to grab the pennies, and pull the pennies into the saw cuts to maintain them at a fixed distance apart. The assembly is shown in Figure H. The other ends of the patch cords will grip jumper wires that plug into your circuit.

You can rest the penny assembly across any small cup that is filled almost to the brim with liquid. I used some little plastic containers like the one shown in Figure I.

Prepare some samples consisting of distilled water, municipal water from your faucet, and bottled spring water (which often contains minerals). You can also try dissolving table salt in water. Label your samples so you don't get them mixed up.

After you note the result using one liquid, remove the pennies, wipe them with a tissue, and try the next liquid. When the sound you hear is higher-pitched, this

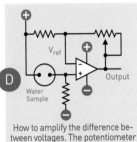

D

How to amplify the difference between voltages. The potentiometer adjusts the gain of the op-amp.

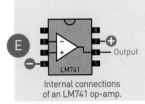

E

Internal connections of an LM741 op-amp.

F

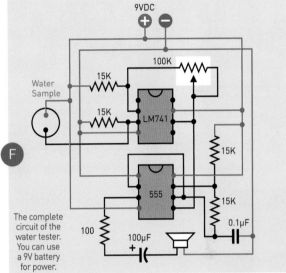

The complete circuit of the water tester. You can use a 9V battery for power.

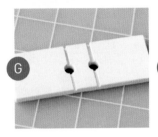

G

H

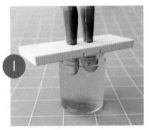

I

means that the resistance is higher and the water is purer.

You can also experiment with other kinds of liquids. I tried mouthwash, milk, and sports drinks (which have a lot of minerals added).

Two factors may make it difficult for you to get consistent results. First, carbon dioxide in the atmosphere quickly dissolves in water, forming carbonic acid. This disassociates to form ions that lower the resistance of your sample. So, keep your container of distilled water sealed, and when you pour some into a cup, test it as quickly as possible.

Second, when you pass DC current between electrodes immersed in a liquid, positive ions gather around the negative electrode while negative ions gather around the positive electrode. This inhibits electric current, so that the resistance seems to increase, and you'll hear the sound from your circuit gradually rising in pitch. You

can minimize this effect by agitating the liquid. Commercial equipment deals with this problem by using AC instead of DC, while running liquid past the electrodes in a steady flow.

Even though you can't expect precision results, you should be able to show that tap water conducts electricity better than distilled water. And if anyone doubts this — just ask them to listen to the difference. ⏻

CHARLES PLATT is the author of *Make: Electronics*, an introductory guide for all ages, its sequel *Make: More Electronics*, and the 3-volume *Encyclopedia of Electronic Components*. His new book, *Make: Tools*, is available now. makershed.com/platt

[+] For technical background on using resistance to determine water purity, check out www.analytical-chemistry.uoc.gr/files/items/6/618/agwgimometria_2.pdf

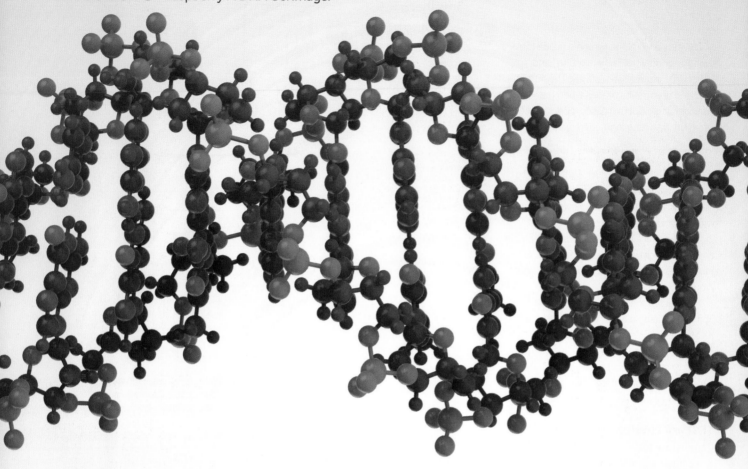

Written by Dr. Lindsay V. Clark

Documenting DNA

Capture your bio-experiments with a Raspberry Pi gel imager

Any genetics lab or DIY biohacker needs to be able to visualize DNA and RNA, and a common technique for doing so is *agarose gel electrophoresis*. The sample is loaded into wells along one edge of the gel, and then a voltage gradient is applied to the gel, attracting DNA and RNA to the anode due to their negative charge. Smaller DNA and RNA fragments migrate through the gel more quickly, resulting in fragments being separated by size.

A dye in the gel binds to the DNA or RNA and fluoresces under ultraviolet light, so we use a UV transilluminator for visualization after the gel is done running. Typically a DNA "ladder" is added to one or more wells, giving a consistent set of bands that the researcher can use for estimating the fragment size of the samples. It's also possible to slice out sections of the gel to

isolate and purify particular DNA fragments.

Our lab has a UV transilluminator, but for taking gel photos we had been reliant on using imagers in other labs. It's slightly inconvenient (and slightly hazardous) to have to walk around the building carrying an ethidium bromide gel, plus if you don't have a key to the other lab you are dependent on their schedule, and I hate to impose. I'd been thinking of trying to make an imager using a Raspberry Pi, so when the imager at one of our go-to labs broke down, I finally did it. I don't know much about optics so I did a bit of research online and found people who had done similar things.

The whole thing cost our lab about $150. A couple caveats: 1) It doesn't zoom or focus. I am OK with this since the camera is positioned such that it can get a decent picture of any gel. If I need a publication-

quality image, I'll consider other options. 2) I just used a styrofoam box rather than buying or constructing something fancier. We already have the UV transilluminator in a separate room with UV face shields available. If you want to have this setup out in the open in your lab, you might need a box that will completely contain the light from the transilluminator.

SET UP THE RASPBERRY PI

If you're new to Raspberry Pi, visit this project online at makezine.com/go/raspberry-pi-gel-imager for step-by-step setup instructions.

Once the Pi and camera are set up, take some test pictures of a printed sheet of paper by typing:

```
raspistill -o ~/Desktop/test.jpg
```

to see how far away you will want the

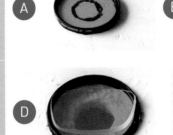

TIME REQUIRED:
1 Hour

DIFFICULTY:
Easy

COST:
$150

MATERIALS

» **Raspberry Pi board, camera, power supply, and SD card with operating system** I find it convenient to just buy a kit, which includes a case for the Pi and costs just under $100.
» **Computer monitor** for programming the Raspberry Pi and viewing images
» **USB keyboard and mouse**
» **Cheap pair of reading glasses** I bought +2.00 glasses from the pharmacy, although since the focus isn't perfect I wonder if a stronger pair would be a little better. The Raspberry Pi camera has a fixed plane of focus from 1m to infinity. The reading glasses bring that plane of focus a bit closer.
» **Orange camera filter** This is not strictly necessary but will filter out the glare from the UV lamp and make the picture much better. And I can use it on the lab's DSLR camera for a publication-quality image. This is for ethidium bromide gels; if you're working with a dye that emits a shorter wavelength, orange might not be the right choice.
» **Styrofoam box** that can fit over your gels, and is about 6"–12" deep in its outer dimensions. Eventually I found the focus was better if I lifted the box up so that my camera was about 12" away from the transilluminator.

TOOLS
» **Screwdriver**
» **Utility knife**
» **Paper clips**
» **Duct or packing tape**
» **Aluminum foil**

DR. LINDSAY V. CLARK is a plant geneticist studying bioenergy grasses in the Department of Crop Sciences at the University of Illinois, Urbana-Champaign. She also creates software for analyzing genetic marker data in polyploid organisms, and teaches R programming.

camera to be from your gel (i.e., make sure you have the right size styrofoam box). Maybe get someone to lend you a second pair of hands, holding the reading glasses in front of the camera while you snap some pictures. The part of the camera board where the cable goes in corresponds to the bottom of the resulting photos, although you can always rotate photos later.

BUILD THE IMAGER
CUT BOX FOR COMPONENTS

Flip the styrofoam box upside down. In the center, trace around the camera filter. Then draw a smaller circle within for the "aperture" (Figure A).

Cut the aperture hole straight through the box (Figure B). Widen the aperture as much as you can while still leaving a rim to hold the filter.

Carve out a shallow circle to seat the filter (Figure C) — you may even want it to fan out in a cone shape underneath.

Pop a lens out of the reading glasses and place it atop the filter (Figure D).

Alternatively, you could position the pair of glasses so the center of one lens is right in the center of the filter, then trace around the glasses. Carve out a shallow trench that the glasses will sit in — they'll need to be immediately next to the Pi camera.

MOUNT THE CAMERA

Attach the camera to the Pi with its cable (Figure E). Unfold a couple of large paper clips, bend them in half, and thread them through the holes in the camera board. Then press the paper clips into the box to position the camera right above the lens (Figure F). Use duct or packing tape to

secure the filter, lens, and Pi (Figure G), making sure not to cover any port you need.

I took a few test pictures here to make sure I was happy with it, and also widened the aperture a bit more at this point. Then I taped aluminum foil over the assembly to block out ambient light. (Figure H). Figure I shows my final setup on top of the transilluminator.

SNAP YOUR SAMPLES

You're ready to take pictures of actual gels! By default, **raspistill** shows a preview for a couple of seconds before taking the picture, in which time I can shift the box a little bit in order to line it up right. Figure J shows the final product.

Just like a regular computer, you can stick a USB flash drive into the Raspberry Pi to retrieve your image files. ◕

Get Started with
In-Circuit
Debugging

Use a debugger to spy on your microcontroller code while it's working

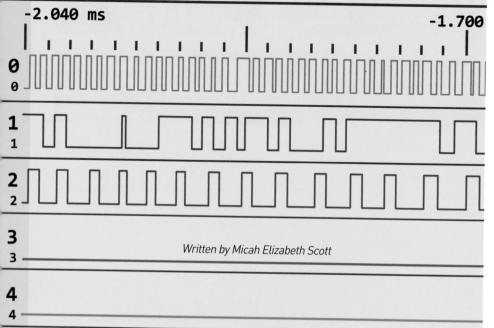

-2.040 ms   -1.700

0
0

1
1

2
2

3
3

4
4

Written by Micah Elizabeth Scott

MICAH ELIZABETH SCOTT is a computer engineer and artist who makes videos about electronics and reverse engineering.

IF YOU'RE MAKING (OR DISASSEMBLING) ANYTHING COMPLICATED, YOU MIGHT YEARN FOR A BETTER VIEW INSIDE.
An oscilloscope or logic analyzer can be a vital tool for digital electronics, and also surprisingly useful for embedded software. Your code can help too, via messages logged to file or serial port. But sometimes you really need an interactive view of your program's internals, and on an embedded system this means you need an *in-circuit debugger (ICD)*.

Some of you already know about source-level debugging from other contexts, like debugging desktop applications in Xcode or Visual Studio, using breakpoints and single-step functionality for scripts in a web browser's console, or using a standalone debugger like the GNU Debugger, WinDbg, or LLDB. If you're more used to Arduino, however, this kind of tool might be new to you.

Debugger generally refers to a system made up of debugging software on your PC, possibly software on the chip you're testing, and usually some hardware both inside and outside the chip. Many microcontrollers have an *on-chip debugger (OCD)* baked into the silicon.

A debugger can step through code one line or instruction at a time while showing the contents of variables, and can view or edit memory. You can run your program at full speed until it hits a *breakpoint* — an intentional pause in your code — where it stops and the debugger resumes. With a debugger attached, you can also typically interact with a processor's peripherals, load programs into flash memory, and read the flash contents (unless protections are in place to prevent this!).

What's a Debug Server?

The debugger app on your PC keeps a detailed memory map for your code, but you also need some hardware — or software — on your microcontroller to provide for reading and writing memory, trapping breakpoints, and running instructions in a controlled way. One solution is a *debug server*.

The *GNU Debugger (GDB)* defines an especially simple server protocol for use over a serial port or network, including `localhost`. Implementations of this `gdbserver` exist for different operating systems. Traditionally this was a software-only component, but the protocol is now commonly used as a gateway to hardware devices or emulators.

Likewise, your IDE options now range far beyond a basic GDB command line, with tools like Eclipse, Visual Studio Code, and IDA Pro supporting the same GDB protocol.

Hardware Debug Ports

It would be convenient if the debug features built into silicon were directly compatible with the GDB protocol, but OCDs are optimized for minimum cost and impact on the overall processor design. Typically debug will be provided via the same port used for programming flash memory. Some chips use vendor-specific protocols, but two industry standards are worth knowing about: *JTAG* and *SWD*.

Whatever the protocol, some hardware is needed to convert back into USB. A wide range of adapters can be used as a debug server via the open source *OpenOCD software*, even hookups you may already have like Raspberry Pi GPIOs or an FTDI serial breakout. And the Black Magic Probe (see "Bug Squashers," right) is an open hardware device that implements the debug server in firmware, providing a virtual serial port you attach directly to GDB.

JTAG: The original standard — It's not a good name. Like JPEG, it doesn't say what the standard does, just who designed it: the Joint Test Action Group. It was designed in the mid- to late 1980s and standardized in 1990, in response to complex circuit board assemblies too difficult to automatically test.

Electrically, the *JTAG standard (IEEE 1149.1)* is a series of shift registers that you can daisy-chain between devices. You might attach your debugger directly to a single device, or to a chain of devices within a single chip or across multiple chips. JTAG looks superficially like SPI, with a shared clock and one-way data input and output pins. But then you see a Test Mode Select

(TMS) pin. Is it a chip-select? No. This is where JTAG starts to get especially low-level. It's actually a bit pattern that drives a *state machine* specified by the standard, which chip makers build upon to create their own JTAG state machines.

The JTAG standard specifies states for selecting a device and reading its 32-bit ID code; and JTAG's *boundary scan* protocol addresses pins on an IC for electrical testing of assembled PCBs. Beyond this, it gets device-specific: FPGAs and processors and memories each have their own protocols. This fragmentation mirrors the fragmentation you see in the whole embedded tools space!

With modern ARM processors, at least, the standards do us a favor. The ARM Debug Interface specification describes a standard *JTAG Debug Port* with a way to access memory, peripherals, and CPU state. From there, memory-mapped registers can modify breakpoints and control the CPU.

SWD: The newcomer — When ARM standardized a way of accessing memory and CPU debug features over JTAG, they also took the opportunity to develop a new alternative protocol: *Serial Wire Debug (SWD)*, which uses a single bidirectional data pin and a modernized packet structure. The reduced pin count makes SWD ideal for smaller embedded processors like the popular ARM Cortex-M series, and it can share pins with JTAG when the processor supports both options.

That's about all you need to know about SWD itself, if you only plan to wire up the debug port on a processor and run high-level tools like GDB. The details of configuring OpenOCD, GDB, and your specific debug adapter will differ by platform, so you'll want to find the configuration file included with OpenOCD that most closely matches your situation as a starting point.

If you want to understand how debugging and even how processors work at a deeper level, the debug port is an excellent place to start digging around! ◉

» For more on ARM's debugging: static.docs.arm.com/ihi0031/c/IHI0031C_debug_interface_as.pdf
» I wrote a simple web-based SWD memory browser for the ESP8266 (github.com/scanlime/esp8266-arm-swd) and an article in *PoC∥GTFO* 10.5 (archive.org/stream/pocorgtfo10#page/n25/mode/2up).
» Find a more complete open source implementation of SWD, try the Black Magic Probe (see "Bug Squashers," right) or Free-DAP (github.com/ataradov/free-dap).

BUG SQUASHERS
Written by Hep Svadja

With the rise of IoT and embedded devices, information security has never been more important. We trust these devices to track our habits, manage our data, move our money, and watch us sleep. Check out some of the tools that specialists use for debugging, and incorporate them into your own hardware development arsenal. Some even provide a DIY BOM, so you can leverage your SMD soldering skills while leveling up your signal analyzing knowledge.

Bus Pirate v3.6
The Bus Pirate (adafruit.com/product/237) is a long-beloved universal bus interface that can talk to most chips via terminal. The PIC24FJ64 processor allows easy firmware updates, and the BP's active community continues to extend usage with a wide array of supported protocols including JTAG, serial, MIDI, PIC, and ARM. The BPv3.6 also includes a binary access mode that can be used with a variety of scripting languages including C, Python, Perl, and more.
BOM: dangerousprototypes.com/blog/2012/03/22/bus-pirate-v3-5a-soic-bom

JTAGulator
Designed around the Parallax Propeller 8 processor, the JTAGulator (grandideastudio.com/jtagulator) allows on-chip access with 24 I/O channels that include voltage input protection circuitry to protect all connected devices. The board includes level translation and voltage filtering with adjustable target voltage from 1.2V to 3.3V, and the USB interface provides power plus onboard terminal access. Supported target interfaces include JTAG/IEEE 1149.1 and UART/asynchronous serial.
BOM: grandideastudio.com/wp-content/uploads/jtagulator_bom.pdf

Black Magic Probe Mini V2.1
If you regularly develop in the ARM Cortex space, you need to get your hands on a Black Magic Probe (1bitsquared.de/products/black-magic-probe). A single device that does both JTAG and SWD, the BMP really shines in its onboard GNU Debugger which does away with the need for intermediary programs while providing full debugging functionality. The board also includes semihosting host I/O support and can receive TRACESWO diagnostics when in SWD mode.
FIRMWARE: github.com/blacksphere/blackmagic/wiki/Debugger-Hardware

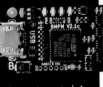

Hep Svadja, 1BitSquared

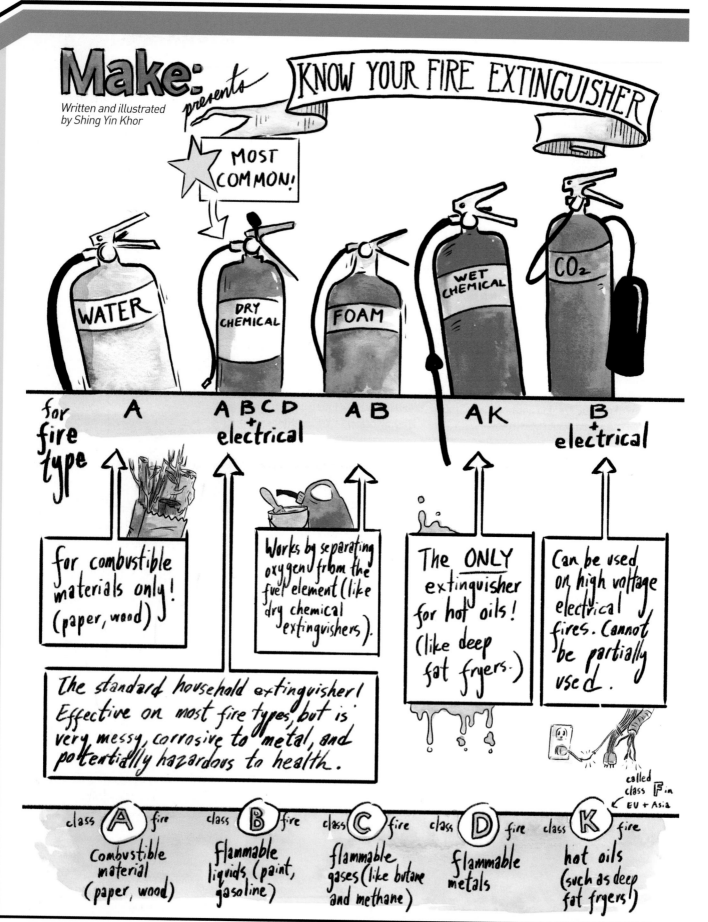

Make: presents KNOW YOUR FIRE EXTINGUISHER

Written and illustrated by Shing Yin Khor

MOST COMMON!

WATER • DRY CHEMICAL • FOAM • WET CHEMICAL • CO₂

for fire type

A | ABCD + electrical | AB | AK | B + electrical

for combustible materials only! (paper, wood)

Works by separating oxygen from the fuel element (like dry chemical extinguishers).

The ONLY extinguisher for hot oils! (like deep fat fryers.)

Can be used on high voltage electrical fires. Cannot be partially used.

The standard household extinguisher! Effective on most fire types, but is very messy, corrosive to metal, and potentially hazardous to health.

called class F in EU + Asia

class Ⓐ fire — Combustible material (paper, wood)

class Ⓑ fire — Flammable liquids (paint, gasoline)

class Ⓒ fire — flammable gases (like butane and methane)

class Ⓓ fire — flammable metals

class Ⓚ fire — hot oils (such as deep fat fryers!)

BOLDPORT CLUB

$30/month boldport.club

Kits are fantastic as a new maker, but once you master hardware it can be hard to find one that stretches your skill set the way your first MintyBoost did. What's an electronics expert to do?

The Boldport Club is a subscription membership from circuit design artiste Dr. Saar Drimer. Each month brings a new soldering challenge, such as the Tiny Engineer Superhero Emergency Kit, which makes a functional circuit with the look and feel of snap out components. Or try the Cordwood puzzles, where you need to figure out the correct layout for the circuit to work. Some kits are useful items, such as Spoolt, that becomes an elegant soldered PCB spool holder when complete.

Each kit contains all the components needed, as well as detailed instructions both on- and offline. Extra kits are available in the shop for club members to grab more of their favorites, but once out of stock they are gone forever! –*Hep Svadja*

SILHOUETTE MINT

$130 silhouetteamerica.com

The Silhouette Mint makes creating stamps fast and easy. Whether you are a makerspace, crafter, teacher, or business owner, making your own custom stamps is a very useful ability. I've used many other methods in the past to make Hackerspace Passport stamps and nothing is as easy as the Mint.

The heavy lifting happens in the Mint Studio companion app. Select a stamp size from the list of available blanks, then import an image. Mint Studio even has filters to turn your full color images into single color stamps. Once your design is finished, load a stamp blank into the Mint and send your image to the device. In a few seconds you will have your brand new stamp! The stamp can then be added to a base and loaded with ink. A fully loaded stamp should last for around 50 stamps before ink must be reapplied.

At $130, you have to really like stamps to justify the cost of this (and of the blanks), but I recently found the Mint on Amazon for substantially less, which was enough to get me to give it a go. This is a bit of a one trick pony of DigiFab tools but one that does its job very well. –*Matt Stultz*

TINY CIRCUITS TINY ARCADE

$60 tinycircuits.com

Retro gamers rejoice! If building an arcade cabinet is one of the many things on your to-do list, but time, money, or space poses an issue, Tiny Circuits has you covered with the quick and easy Tiny Arcade kit.

This kit contains everything you need to build your very own working arcade cabinet in the palm of your hand, including a vivid OLED screen, joystick, buttons, built-in speaker, and rechargeable battery, complete with an acrylic enclosure that fits together like a 3D jigsaw puzzle.

Everything is pre-soldered and takes roughly 10 minutes to assemble. It's also surprisingly ergonomic and comfortable to play. Tiny Arcade comes preloaded with three games: Tiny Shooter, Tinytris, and Flappy Birdz.

With an SD card you can load more games and videos, a few of which can be downloaded directly from Tiny Circuits' website. Best of all, Tiny Circuits also provides an online guide on developing your own video games written entirely in the Arduino IDE.

So if you're looking for a quick and easy — but incredibly cool — kit to build, the Tiny Arcade is a solid project for makers at any level to experience the joy of retro gaming.

–Jun Shéna

CHIBITRONICS LOVE TO CODE CREATIVE CODING KIT

$85 chibitronics.com

Many beginner kits focus on electronic components, which can be frustrating to those new to both hardware and software. Chibitronics uses a papercraft storybook to put the focus on programming. Beautiful copper tape circuit designs take the place of the usual breadboard to teach electronics with an engaging story that will resonate with novice makers of all ages.

Choose either the block programming environment MakeCode or the Chibitronics web interface for Arduino-like programming.

The Chibi Chip microcontroller has conductive pads, which clamp the edge of pages to connect with copper tape traces. There's no soldering required making this perfect for younger makers. *–Hep Svadja*

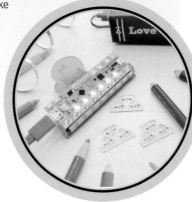

OSKITONE OKAY SYNTH DIY KIT

$50 electronics only/$90 with printed parts oskitone.com

Oskitone makes fun project kits that combine DIY electronics with 3D printing to make synths and other musical instruments.

The OKAY Synth comes with a built-in amplifier, speaker, ¼" audio out jack, six selectable octaves, and one octave of keys (though version two adds a second octave of keys). You can also choose between a complete kit with 3D printed parts or just the electronics and print your own enclosure. Instructions are thorough and you'll learn about the circuit as you build and customize it. All you need is a soldering iron and you're ready to make your own synthesizer.

Speaking of customizing, I also recommend Oskitone's amazing web app at build.oskitone.com where you can tailor the whole build to your liking! What is greater than a DIY kit where the finished product inspires even more creativity? *–Andrew Stott*

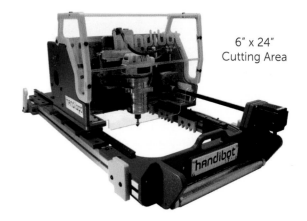

SHOW&TELL

If you'd like to see your project in a future issue of *Make:* magazine submit your work to makershare.com/missions/mission-make!

1 When challenged to design and build a piece of furniture out of a single piece of plywood, **Jeffrey Burke** saw it as an opportunity to create a coffee table to give his mom for Mother's Day. With drawers that open from both the front and back for easy access, a slick reclaimed barn wood edge banding, and a host of finishing applications, it's easy to see the love that went into this piece. makershare.com/projects/coffee-table-mothers-day

2 The modernist architectural design influence of Ludwig Mies van der Rohe is evident in the sharp edges and stylistic choices of this Bluetooth speaker. Maker **Jeffrey Burke** used the project to learn to solder, and with a sub-$50 price tag it makes an excellent point of entry to learn woodworking, joinery, and electronics on a reasonable budget. makershare.com/projects/diy-bluetooth-speaker

3 On a quest to help students memorize their multiplication tables, **Jo:ha Dreyer** aimed to demonstrate how multiplication directly translates to area calculation. This passion began while tutoring others in math, and was further fueled through helping his son learn Arduino and Raspberry Pi projects. From there it was a short step to creating math toys and eventually this 144-LED multiplication machine that can quickly visualize area. makershare.com/projects/matrix-math-machine

4 This is a story of necessity. **Andre Ferreira** had a nice workshop, but his tools were all over the place, so he decided to create a toolbox so he'd have more working space. With found plywood and reclaimed drawer slides, the cost of this entire project was just time. The final build is a beautifully crafted, personalized tool chest. makershare.com/projects/toolbox-reclaimed-materials

[+] **Read about our Editors' Choice, RoboRuckus, on page 58.**

Matrix Math Machine